On the concept of time and other accidents

Luca Gammaitoni

On the concept of time and other accidents

simple scientific digressions for curious people

Acknowledgements

The author, a university professor for more than three decades, is deeply indebted to several generations of students he has met in his business, as well as to the many colleagues and collaborators with whom he has had the fortune to work over the years.

Special thanks go to Adi R. Bulsara for a careful reading of the first essay entitled *on the inevitability of time and the fundamental role of fluctuations* and to Ester Pascolini for her continuous support and encouragement.

Contents

Introduction (p. 9)

1. On the inevitability of time and the fundamental role of fluctuations (p. 10)
2. Where has my parking gone? (p. 22)
3. What will I do with my garage? (p. 25)
4. Predicting the future of the epidemic (p. 29)
5. Excuse me, may I trace you? (p. 33)
6. Who's afraid of 5G (p. 37)
7. Post Covid-19: the world to come (p. 41)
8. How intelligent is Artificial Intelligence? (p. 45)
9. As to why it's impossible to find something if you don't know what you're looking for (p. 49)
10. Is man a machine? (p. 54)
11. Black hole is not a hole (p. 58)
12. Computers as small as specks of dust (p. 61)
13. Waves of epidemic (p. 65)
14. Unlikely swabs, effective vaccines and other accidents (p. 69)
15. Much noise about nothing? (p. 73)
16. The noise you don't expect (p. 77)
17. Beyond the Standard Model, in search of a New Physics (p. 81)
18. Vibrations and that Swiss invention of long ago (p. 85)
19. Atoms and bits (p. 89)
20. Memory doesn't last forever (p. 93)
21. If data is the new oil, who are the new oilmen? (p. 98)
22. Surveillance Capitalism and the future technology (p. 101)
23. Physics: A quality Nobel (p. 105)
24. The Omicron variant and the evolution of the pandemic (p. 108)

Brief scientific biography of the author (p. 112)

Introduction

This book presents 24 short pieces on various scientific topics: from the energy of the future to artificial intelligence, from microcomputers to the role of fluctuations. From self-driving cars to the laws that rule epidemics. 23 of them have already appeared in Italian, in the magazine "L'Osservatore" in the monthly column "Tracce di futuro", in the year 2020 and 2021.

The first piece, which opens the book, is instead unpublished and is about the concept of time and its interpretation in physics.

1. On the inevitability of time and the fundamental role of fluctuations

There are summer afternoons and winter evenings in which time is, apparently, still.

In the summer, if you are alone and look towards the blue sky, it seems that nothing changes. The few clouds on the horizon are always in the same position. Yet, if we take a photo with our mobile phone and, after having dozed off for a few minutes, wake up and compare the photo with the sky, we would notice that there are small but not entirely negligible differences.

In the winter, alone in the house sometime after the sun has set, an incandescent light bulb (one of the few left on the market) shines on us. Its light is constant and warm. It spreads a sense of peace and tranquillity in the quiet room. Yet, if we could take high repetition photographs, we would notice that the room lighting tends to dim and then come back on again, fifty times every second.

These are two examples that help us understand our relationship with time. We can say that when we do not perceive any variation around us, time does not seem to pass. Yet we know this is a deception. In fact, it is sufficient to use some technological aid to discover that time has continued to flow, in the oblivion of our senses.

But, it is worth asking: if there were a situation in which, even with the use of the most sophisticated tools, we were unable to identify any changes, could we then still say that time exists?

I do not think so. A time that does not pass is a time that does not exist. Who needs the concept of time if nothing changes? Aristotle must, also, have thought the same thing; he wrote twenty-four centuries ago:

. . . However, time is not even without change: if, in fact, there is no change in our mind or we do not realize that we are changing, it does not seem that time has passed, just as it happens upon awakening to those of whom it is said that in Sardis they slept with the heroes: they, in fact, connect the posterior hour to the anterior one and make them one, eliminating what is in the middle due to the fact of not having perceived it. Just as, therefore, if the hour were not different, but were one and the same, there would be no time, so even in the case in which one does not realize that it is different, it does not seem that there is time in between.

Aristotle, Physics, IV II

The idea that time exists only in the presence of a change is the basis of what in philosophy are called *relational theories of time*. Perhaps the position of Parmenides, which supported the illusory nature of change and therefore of time, can be ascribed to this trend.

The so-called *substantive theories of time* are opposed to the *relational theories of time*, according to which time exists even in the absence of any event and therefore of something that changes. The origin of this absolute conception of time is usually traced back to Newton.

Absolute, true and mathematical time, by itself and by its own nature, flows uniformly without relation to anything external, and by another name it is called duration: relative, apparent and common time is a reasonable and external measure (whether accurate or non-uniform) of duration by means of movement, which is commonly used instead of true time, such as an hour, a month, a year.

I. Newton, Philosophiae Naturalis Principia Mathematica

It goes without saying that Leibnitz, a contender with Newton of the birth right of the invention of the infinitesimal calculus, supported the opposite thesis, namely the *relational* one. Also Hume who, in 1739, wrote:

The ideas of space and time are therefore not separate or distinct ideas, but simply those of the way or order in which objects exist: or in other words, it is impossible to conceive of a void or an extension without matter, or a time, when there is no succession or change in any real existence.

D. Hume, A Treatise of Human Nature

The name of Albert Einstein must certainly also be associated with the *relational* party; he, to avoid the embarrassing questions about what time was, took refuge in the concept of measurement (of intervals) of time, achieved through physical systems that change periodically and regularly: clocks.

It is thanks to Einstein that today we know that the cyclic tickling of clocks occurs in a different way, if we compare two that are in motion one with respect to the other, or in the presence of a gravitational field much larger for one than for the other. In this sense, Einstein's great contribution does not add or detract from the idea that time and change are intimately correlated, instead it determines the way they flow in different physical conditions and, for the first time since Galileo's time, shows how time and space are actually components of a single physical quantity that we call space-time. We will come back to this concept shortly.

Analogously to the different views of time between *relationalists* and *substantialists*, there are also two other philosophical views: *presentism* and *eternism*. *Presentists* affirm that only the present instant is real while the future and the past do not yet exist and no longer exist. Conversely, the *eternists* have a conception of time more similar to space: the present corresponds to a temporal place (here and now) while past and future are other equally real temporal places. Clearly, time travel enthusiasts can only join the party of the *eternists*, otherwise they would not have a destination to land at in the unlikely event that a time machine was invented.

In the opening chapter of the book *Perchè è difficile predire il futuro* (Why it is difficult to predict the future), written together with my colleague Angelo Vulpiani, we briefly talked about which conception of time can be adopted in order to talk about future forecasting. I will not repeat here the considerations made there, because they are not relevant to what I am interested in saying now; however, if you wish, you can consult that chapter to deepen the past-present-future sequence matter.

Vacuum does not exist and immobility is prohibited

In not so recent years, physics research has added an important piece to the mosaic of nature, making a decisive contribution to the debate between *relational theories* and *substantial theories* of time. It has clearly established that in nature there cannot be emptiness and there cannot be stillness.

The author of this fundamental result was Werner Heisenberg and his uncertainty principle, currently the most fundamental principle we have in physics, is the basis of the most complete theory of motion available to date: quantum mechanics.

To see what the uncertainty principle is all about, we must recall the heart of the dispute between the two opposing conceptions of time. According to the *relationalists*, in the absence of movement there would be no time. This situation would necessarily occur in an empty space. For the *substantialists*, on the other hand, time would continue to exist (and to flow) even in an empty space. Quantum mechanics intervenes in this debate simply by explaining that there cannot be an empty space. In fact, thanks to Heisenberg's uncertainty principle, even in a region of space apparently devoid of matter and energy, in short, devoid of what is called a *field* in Physics, we are witnessing the continuous creation and annihilation of pairs of particles. These pairs consist of a particle and its antiparticle, for example electron-positron, which continually appear and disappear without stopping.

According to this model, the vacuum of quantum mechanics is actually a space characterized by a teeming of events which, on a purely random basis, take place continuously: creation and annihilation follow one another unpredictably, without end. These particle-antiparticle pairs can only exist for a short time interval Dt. Incredible to say, but actually verified by numerous experiments, this pair of particles is born out of nowhere, momentarily violating the conservation of energy, for a quantity DE which is required to bring these particles into existence. All this, astonishing as it may be, is absolutely compatible with the known laws of physics, provided that the product of these two quantities remains very small. The time-energy uncertainty principle can be interpreted by saying that there can exist two particles whose total energy is DE as long as their existence does not last more than a time Dt, inversely proportional to DE. The constant of proportionality is linked to a quantity, often indicated by the letter h, identified as Plank's constant (from the name of the physicist who first identified it), and has a very small numerical value. To give you an example, an electron-positron pair, which typically has an energy on the order of 10^{-13} Joules, can exist for a time on the order of 10^{-21} seconds: very little but, nevertheless, a non-zero time interval.

As we said, the effects of the uncertainty principle are not limited to excluding the existence of the void but also extend to the *prohibition of immobility*.

This time the two physical quantities involved are not time (t) and energy (E) but position (x) and momentum (p) (momentum is simply the product of mass times the speed of a body). In this case Dx represents the uncertainty in the position of a material body and Dp the uncertainty in its momentum.

The Heisenberg uncertainty principle, expressed in terms of Dx and Dp, is interpreted as follows: if we make a measurement, then it is not possible to know both the position and the speed of an object at the same time with absolute precision (i.e. with zero uncertainty).

This prescription has important consequences on the concept of immobility. Let's see why.

Let's start with an example. Suppose we place a ball on a horizontal table and know its position with an uncertainty Dx, then, due to the Heisenberg relation, we will have a minimum uncertainty about its speed Dv which is inversely proportional to the uncertainty about the position, meaning: the smaller the uncertainty about the position (i.e. the more precisely we are able to measure the position), the larger is the uncertainty about the velocity (we will be less able to measure its velocity at the same time).

Measuring something is the activity that physicists have always done; for example, measuring the length of a table or the speed of a bullet. A given uncertainty is, necessarily, associated with each experimental measure. Let's say that the length of a table is 1.90 m with an uncertainty of plus or minus 1 cm. This "plus or minus" constitutes our uncertainty Dx. Generally, if we change the measurement instrument (the meter) and choose one with a higher resolution (a stick with millimeter or tenths of a millimeter marks) we can improve our measurement and reduce uncertainty. The same is true for speed measurement which is based on length and time measurements.

However, Heisenberg's principle tells us that, even if we continue to refine our measuring instruments and choose better and better ones, we will arrive at a point where we can no longer reduce the uncertainty on both of these quantities (position and velocity) at the same time.

Why is there this insurmountable limit to the accuracy of our measurements?

The explanation of this strange behavior of nature is that, in reality, what appears to us to be a material object (a ball) is a more complicated thing, whose behavior is not quite that of a ball that we have known since childhood, but a kind of middle ground between a ball and a wave of the sea. I agree that said like this, it might sound crazy but, believe me, there are few better explanations than this.

For a sea wave, the concept of position is certainly more vague than for a small ball. In fact, we cannot say that the wave is in a given point but we can at the most say that the wave occupies a certain region of space, much larger than a point. This vagueness of localization is reflected in our measurements and defines the uncertainty that characterizes them.

Due to the impossibility of reducing the uncertainty about the position to zero, our ball will not be able to stay still.

To understand this point, let's see what would happen to the ball in the case of classical mechanics (that of Galileo and Newton) and then let's see the same thing in the case of quantum mechanics.

In classical mechanics, the ball on the horizontal table, in the absence of forces, has a position $x(t)$ that changes linearly with time, i.e. the more time passes, the further it moves away from the initial position $x(0)$ with a given initial speed v. Since we are in the absence of external forces acting on the ball, the speed remains constant. If the speed is zero $(v=0)$ then $x(t)=x(0)$ and the ball always remains in the same place. The same thing applies to its uncertainty Dx. If the uncertainty on the speed Dv is zero, then $Dx(t)=Dx(0)$ and the uncertainty does not change over time. In conclusion we can say that the ball is on the table, in a position $x(0)$, with an uncertainty $Dx(0)$ and it remains forever in that area.

The same is not true in quantum mechanics. Here the position of the ball at successive instants of time is described by a more complicated equation than that of classical mechanics, discovered by Erwin

Schrödinger in 1925. In this case, in the absence of external forces, the position of the ball remains substantially centered around the initial value, with an uncertainty Dx which however inevitably grows over time. In fact, due to the existence of the uncertainty principle, this time we cannot have (at the same time) $Dx(0)$ and $Dv(0)$ as small as we like and therefore, for each finite value of $Dx(0)$, we will have a corresponding value of $Dv(0)$ resulting in a necessary change of $Dx(t)$ over time (see appendix for details).

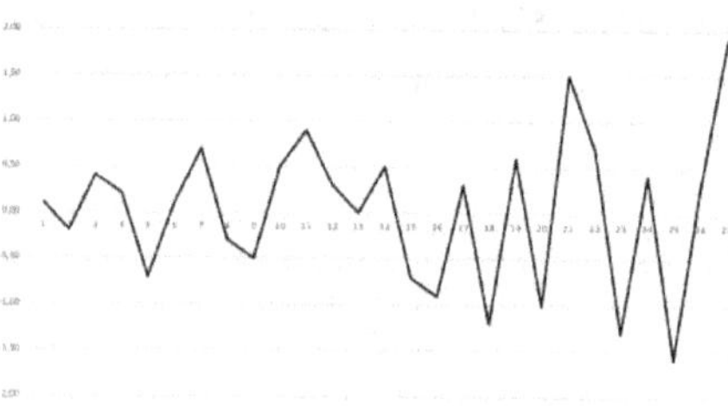

Figure 1

The fact that this uncertainty inevitably grows over time, brings with it the idea that immobility is prohibited. In fact, if we carry out a series of measurements of the position of the ball, spaced by a certain interval of time, e.g. one second, we will find that the position of the ball varies from time to time, describing a set of values whose average distance from the initial position will increase as time increases. Therefore, the position of the ball is not constant, indicating that the ball itself has moved after each measure, moving (on average) further and further away, from its starting position.

But there is more. The position of the ball will not follow a regular trajectory but, connecting the points associated with successive measured positions with straight lines, we will find a very irregular trace, similar to the one sketched in Figure 1.

Upon closer inspection of this zigzagging process, a phenomenon well known to physicists since the early nineteenth century comes to mind: Brownian motion, that is the motion of a pollen grain on the surface of a fluid. This type of motion has been studied for a long time and, once again, its most genuine interpretation came from Albert Einstein who, in 1905 wrote an important article to predict how the uncertainty of the position of the pollen grain evolves with time. This is a typical phenomenon of the physics of fluctuations, where the position varies randomly thanks to the myriad of shocks to which the grain is subjected, due to the presence of water molecules that continuously stir due to the finite temperature of the liquid.

Random forces, fluctuations, Brownian motion. What do these pillars of classical statistical mechanics have to do with Heisenberg's uncertainty principle and quantum mechanics? There is no doubt that there is at least one phenomenological analogy: the quantum particle changes its position randomly with an uncertainty that grows over time. The position of the pollen grain also changes its position in a causal way with an uncertainty that grows over time. In this last case, however, the cause of this original behavior, as the Scottish botanist Robert Brown discovered in 1827, is the presence of a very large number of water molecules, invisible to the microscope, whose frequent and unstoppable collisions produce the random motion of the pollen grain. The analogy with Brownian motion would lead us to conclude that a similar cause might be the reason of the behavior of the position of our quantum particle, however, although many attempts have been made to identify this cause, we are not presently able to say which one it is.

Space-time quanta

The analogy between the thermal fluctuations of Brownian motion and the quantum fluctuations in the position of the particle has attracted attention in the physics community since the middle of the last century, when the Hungarian physicist Imre Fényes, published a work in 1946, in which he tried to interpret the Schrodinger equation as a special form of *diffusion equation*, much similar to the Fokker-Plank equation commonly used for thermal fluctuations. The same approach was then resumed and further developed by Louis De Broglie in the sixties and by Edward Nelson, with the so-called theory of *stochastic mechanics*, in the mid-eighties.

At the bases of these approaches, there is the idea that the presence of fluctuations lurks in the most minute structure of space-time. The same fluctuations would also be at the origin of the incessant process of creation and destruction of particle-antiparticle pairs of which we spoke a little above, when we mentioned the vacuum fluctuations. By physicists, this fluctuating structure of space-time is sometimes referred to as *quantum foam*: a name invented by the American physicist John Archibald Wheeler.

Let's leave *quantum foam* for a moment and briefly return to the space-time introduced by Einstein. Thanks to the work of the Dutch physicist Hendrik Lorentz, summarized in his famous laws of transformation of lengths and times, in the presence of two reference systems in motion at constant speed with respect to each other, we now know that space and time are actually intimately connected and the modifications of one end up acting on the other as well. A typical example is the phenomenon of time dilation and length contraction, typical of the special theory of relativity. But we are digressing, and instead we need to stay focused on this fluctuating space-time. Despite Wheeler's suggestive vision, if we were to keep faith with the analogy with Brownian motion, it would remain to determine what actually fluctuates. What does it mean to say that space-time is the seat of incessant fluctuations?

A possible clue could come from the most recent unification theories of physics: *string* theory and *quantum loop* theory. These are two of the most accredited theories to explain how the gravitational force, one of the forces known to man for the longest time, can be made compatible with the other forces that we have discovered later: the electromagnetic force, the nuclear strong force and weak force.

While these last three forces admit a conceptual framework of common description, generically called *Field theory*, the gravitational force appears to show a completely different character because it can deform space-time. In fact, it is known that in the vicinity of large gravitational masses, space can bend and time slow down. This peculiarity that allows the gravitational force to intervene on space and time makes it difficult to treat it in the same conceptual scheme as the other three forces.

The two recent theories we mentioned above, however, have a common aspect that could represent an interesting clue for the unification of all forces. It is, as you imagine, the hypothetical presence of fluctuations in the dynamics of entities underlying space-time. In this perspective, the space and time we are used to would be nothing more than *emerging realities* that we can experience at the macroscopic level while, at the microscopic level, there would be quanta of space-time connected to each other by phenomena of *quantum entanglement*. In order to better grasp this point, we might use an analogy: it is as if, sailing with a boat in the sea, we believe that all reality is made of sea water, ignoring that the liquid is a macroscopic state of matter determined by existing relationships between the water molecules, in turn made by closely connected hydrogen and oxygen atoms. The phenomena that we observe: the sea foam, the waves, the sea eddies, are rather to be considered as emerging phenomena of molecular dynamics, rather than realities in their own right.

For now, we do not know if these theories will have the ability to explain new undiscovered phenomena in the future and therefore to be scientifically validated. At the moment they are little more than hypotheses, certainly suggestive and intriguing.

The fact remains that fluctuations, randomness, noise... however you look at it, instead of being a mere disturbance of our perceptions, appear more and more as inevitable elements in our description of reality. Time and fluctuations are therefore two intimately connected concepts. If I had to choose an image to represent this idea, I would certainly choose an hourglass. The instrument that has always been used to measure time whose succession in the fall of sand grains is nothing more than a random sequence of events, dominated by fluctuations.

Appendix: some formulas, just enough.

The most popular formulation of the uncertainty principle is often represented by an equation of this type:

$$Dx \, Dp \geq \frac{h}{4\pi}$$

The two physical quantities involved are the position of a particle (x) and its momentum (p) which is nothing more than the product of its mass m by its velocity v such that $p = mv$. In this case Dx represents the uncertainty in the position of a material body and Dp the uncertainty in its momentum as determined following a measurement operation.

A similar relationship also exists for the quantities time (t) and energy (E)

$$Dt \, DE \geq \frac{h}{4\pi}$$

Suppose we place a ball on a horizontal table and its position is known with an uncertainty Dx, then, due to the Heisenberg relation, we will have a minimum uncertainty about its speed which is given by:

$$Dv = \frac{h}{4\pi \, m \, Dx}$$

In classical mechanics, the ball above the horizontal table, in the absence of forces, has an $x(t)$ position that changes with time, according to the equation $x(t) = x(0) + v \, t$. If the speed is zero $(v = 0)$

then $x(t) = x(0)$ and the ball always remains in the same place. The same thing goes for its uncertainty:

$$Dx(t) = Dx(0) + Dv\,t$$

If the uncertainty on the speed Dv is zero, then $Dx(t) = Dx(0)$ and the uncertainty does not change. So the ball is in fact on the table, in position $x(0)$, with an uncertainty $Dx(0)$ and in that area it remains.

The same is not true in quantum mechanics. Here the position of the ball at successive instants of time is described by a more complicated equation (discovered by Erwin Schrödinger in 1925) but which nevertheless, in the absence of external forces, as in our case, tells us that the position remains substantially centered around the initial value, with an uncertainty Dx which however increases over time according to the law:

$$Dx(t) = \sqrt{Dx(0)^2 + Dv^2\,t^2}$$

If for the uncertainty Dx we use the Heisenberg relation above, we get

$$Dx(t) = \sqrt{Dx(0)^2 + \left(\frac{h}{4\pi\,m\,Dx(0)}\right)^2 t^2}$$

which expresses the fact that the uncertainty about the position of our ball inevitably grows with time at a rate depending on how small we can make the initial uncertainty $Dx(0)$.

2. Where has my parking gone?

If I had to choose a slogan to launch a mission to redeem contemporary society, I would choose: "think the future to build the present".

It is all too evident, in fact, that contemporary politics is almost exclusively focused on the management of the existent. Dramatically unable to offer a vision of the future that arouses any hope of change for the better. To convince you, I would like to report a recent conversation I had with the mayor of a town characterized by a beautiful medieval structure. The mayor complained about the stalemate in the works in the city council due to the contrast between those who would like to build a large underground parking lot, thus subjecting the municipal coffers to considerable stress and those who, instead, would like to build three distinct parking spaces on the surface, ending up sacrificing green and pedestrian areas. What decision to make?

My answer was: neither. The problem of parking in medieval villages, splendid architectural jewels, is a problem that will disappear on its own in a few years, ten at the most.

The mayor looked at me like I was crazy. Thus, I tried to explain what, in my opinion, will happen in the road transport sector in the near future. The novelty, unknown to most of the local politicians, but well known instead to technicians and economists from all over the world, is that self-driving cars are coming and these will completely revolutionize our way of life.

For some years now, various car manufacturers have been developing prototypes of cars which, thanks to a complex system of sensors and computers, are able to drive themselves in the most disparate situations: from chaotic city traffic to lonely country roads. The tests are at a very advanced stage and already today it can happen that we come across the first prototypes on our roads. It is reasonable to think that within a few years we will have our streets flooded with these types of vehicles, fully fulfilling the promise originally inherent in the term "auto-mobile" (self mobile).

An easy scenario to imagine is the following: if I have to go from place A to place B, I can do it simply by typing the two positions on my smartphone and get *immediately* a number of offers from vehicles of different companies that are passing nearby . Having chosen the most convenient service for me, I can wait for the car to arrive and be driven, alone or with other passengers, to my destination, paying conveniently online. When the cars have no requests, they can go by themselves to an area outside the city used for parking.

This scenario is technically possible today and only some administrative and legal constraints have to be released for it to become a daily reality. Let us ask ourselves a few questions about how it will affect our life.

Users. It is reasonable to think that, while a significant percentage of drivers remain who will keep their old-stile drive cars, a growing fraction of users who want an automatic car will gradually make their way. Who will be the first buyers? Elderly people willing to move without the constraints of age? Anxious parents, tired of ferrying underage children from a piano lesson to a dance lesson?

Purchase or rental. Will it still be reasonable to buy a personal car? This will probably have negative consequences on the subcompact market while perhaps the sports and luxury market will strengthen. Is it good for our economy? What will happen to public transport systems such as the train or the subway?

Works that disappear, change, are born. What impact will the presence of self-driving cars have on the taxi economy? And what about car rentals? Will the garage business be impacted? How will the motor liability insurance sector change? What will happen to the truck drivers?

Urban planning. Will parking spaces in the city disappear? Will the phenomenon of unauthorized parking attendants cease? Will there still be cars parked out of place or in a double row? Will cities change their appearance?

Health & Safety. Will the number of accidents decrease? Will health care expenditure change? Will the number of car thefts decrease? In the event of an accident, how will liability be handled?

As we can see, there are many new scenarios and questions to be answered that this *simple* innovation will bring to our society. And this is only one of the many innovations that we will see in the near future.

It is very important that we are aware of the upcoming changes because, willing or not, they will happen. It is up to us to do our best to orient them towards the best. This is a task that falls in particular to politicians, to those who manage our society or who are candidates to govern the future of countries and communities.

Finally, for everyone, a question that I deliberately left last: will it still be necessary/desirable to have a garage next to one's home? And who already has it, what can they do with it in the future since they will no longer have a car?

In the next episode we will start from this question. Prepare yourself.

Appeared in Italian in N. 4/2020 *L'Osservatore magazine*.

3.   What will I do with my garage?

The first time I heard about the *Internet of Energy* was on October 28, 2013. I was invited to present the results of the research conducted in my laboratory at a conference entitled "Berkeley Symposium on Energy Efficient Electronic Systems" and I was reviewing my notes when from the stage, Arun Majumdar, who had recently become head of Google Energy, had uttered those strange words. The idea that Majumdar was talking about was not entirely new to me, but seeing it declined in that way was truly an epiphany.

In a nutshell it is this: do you know how the Internet works? On the Internet, you have a very large number of computers connected in a telecommunications network. Each computer can act as a subject that requests information (web client, for example) or provides it (web server). Why not do the same thing with electricity, where instead of computers there are many small or large power generation plants connected in a current transport network? Large power plants are thermoelectric, hydroelectric or nuclear power plants. Small power plants can be those that use renewable energy, such as solar or wind power.
In recent years this idea has gained momentum and has been enriched with many details and tools, such as the concept of "smart grid": an electricity distribution network equipped with sensors that can allow a more efficient management of the electricity dispatch.

We have not yet reached the *Internet of Energy* but that is the path we have taken. For some years now, in fact, throughout Europe it has been possible to produce electricity in one's home, using solar panels for example, and to introduce the energy that we do not use directly into the public network for distribution to other users.

Let's see how it can work for a normal family that wants to take this path. First of all, an electricity production system must be installed. The simplest and cheapest way is to choose a photovoltaic system to be mounted on your roof. A 3 kW system is sufficient for a family of 3 or 4 people. The kW symbol represents the quantity of 1000 Watts and the Watt is the unit of measurement for *power*.

To understand what *power* is, let's give an example: suppose you have to carry your supermarket groceries up the stairs of your home. Since you live on the first floor let's say that the difference in height you have to cover is about 3.5 meters. If your shopping weighs 3 kg, then to carry out the transport you need an energy equal to about 100 Joules (1 joule is the energy you need to lift an apple one meter). If it takes you 100 seconds to bring up the groceries, then you need the power of 1 W (i.e. one Joule per second, since power is energy divided by time). If you want to go faster, for example in 10 seconds, then you need the power of 10 W (100 Joules / 10 seconds).

A typical family home generally has 3 kW of electrical power and its inhabitants must decide each time how to use it. Do you remember when you were told not to turn on the hairdryer, the washing machine and the electric water heater at the same time? The reason is that the required power would exceed 3 kW and your system would fail because it does not have all that power available.

With a 3 kW photovoltaic system you can have around 13 billion Joules in one year in the latitudes of Switzerland/Northern Italy. A number that seems impressive but which is approximately how much an average family spends in a year: about 9.72 billion Joules, or expressed in kW per hour, 2,700 kWh.

To install your family system, you need about 20 m^2 of roof (preferably south oriented) and it will cost you around 8,000-9,000€ (but prices are still falling and then there are public incentives to be counted on a case-by-case basis).

The problem with solar energy is that the sun isn't always there. It is not seen for at least 12 hours a day and is sometimes in short supply even during the other 12 hours, especially in winter. What do you do then?

You have two paths: 1) you rely on an external supplier (a large company) and make a deal: it takes up the energy you do not use during the day and gives back you the energy you need during the night or when you need it. 2) store the surplus produced electricity in special batteries to be able to use it at a future time.

This second hypothesis, once technically impractical, is becoming more interesting day by day. And it is here that we reconnect to the question with which we left the previous episode of our column: what will I do with my garage? From a perspective of life without your own car, the garage will become the place for storing the surplus energy produced by your photovoltaic system.

Your decisive step towards energy independence passes from here: a nice large battery pack allocated in your garage. Solutions that are not really cheap but already affordable are starting to appear. The most popular is the one offered by Tesla of the famous Elon Musk company, with the "Powerwall" product: practically a cabinet weighing just over 100 kg which, at the modest price of about 10,000 €, let you have an energy autonomy of a couple of days in case of total absence of sun. Another offer is that of Kyocera, a leading solar company that has developed modular batteries with new technologies (Lithium ions with semi-solid electrodes). Or BOSCH, thanks to its experience with hybrid car batteries or SMA with the Sunny Boy Storage system. As we said, the costs are still high but are falling visibly.

In short, the future promises us an energy management that from large producers will become increasingly closer to the domestic environment, with the possibility for us to become energy producers and not just consumers. Just as the Internet has revolutionized the information market, so this new approach promises to revolutionize the energy market. However, one step is still missing for the analogy to be perfect: currently you cannot sell your energy to your neighbor but only to the national energy manager (in Italy the GSE S.p.A of the Ministry of Economy and Finance, in Switzerland the Pronovo Ltd controlled by Swissgrid), which purchases it from you and resells it to others. At the moment it is an important limitation, but it does not mean that things cannot change soon.

Stay tuned.

Appeared in Italian in N. 8/2020 *L'Osservatore magazine*.

4.   Predicting the future of the epidemic

Is it possible to predict the future? A question that we have asked ourselves many times and that, especially in these days when the coronavirus epidemic is raging, becomes persistent.

Well, the answer is yes, but it is generally very difficult.

The first to find a serious way to effectively predict the future were Renaissance scientists such as Nicolò Fontana, known as Tartaglia, and Galileo Galilei. They were struggling with a rather practical prediction problem: if we fire a cannon and the bullet comes out of the mouth with a certain speed and at a certain angle, where will it fall? How soon will it arrive at the target?

The questions we ask ourselves today, five hundred years later, are not very different from these: if today there are N (number of known infected people) in a certain area, how many will there be in 10 days? How many of these will end up in ICU? How many deaths should I expect?

For over five hundred years, the best way to answer these questions has been to make a mathematical model of the phenomenon we want to deal with. Epidemiologists, a recent class of scientists borrowed from physicists, mathematicians and biologists, are currently under pressure to verify the stability of their models regarding the evolution of what has quickly become a pandemic. In this sector, Italian science is making important contributions with the school founded at the ISI of Turin by Alessandro Vespignani (currently the director of the Network Science Institute in Boston). To this school certainly belong Ciro Cattuto (presently the director of the Piedmontese research center and which I was lucky enough to have as my student at the University of Perugia), Vittoria Colizza and Daniela Paolotti, just to name a few I know personally.

What is a mathematical model and how can it help us predict the future of an epidemic? To answer this question, let's take for example the SIR model, about which we hear so much about in these days in the media. Born almost a hundred years ago from the work of two Scottish scientists, Kermack and McKendrick, it owes its name to the initials of the three English words: Susceptibles (Healthy people, susceptible to contagion), Infected (already infected) and Removed (removed because they are healed / immunized or deceased). It is also called the compartment model because it divides the entire population of a certain region into three compartments.

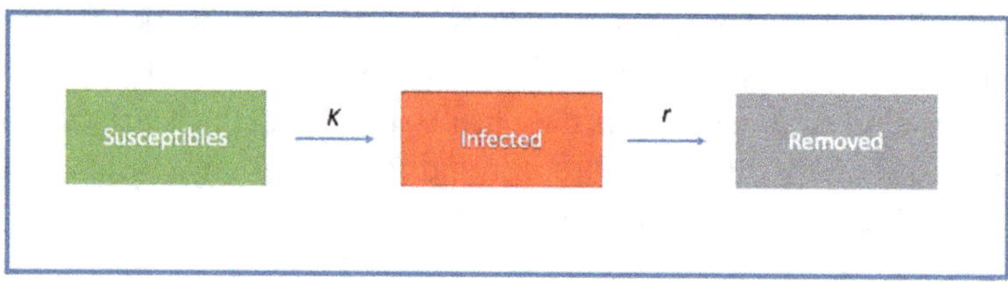

Figure 1

The passage from one compartment to another is governed by a probabilistic rule. For example: the passage from the susceptible compartment to the infected compartment is regulated by a parameter, which we will call K. The value of K tells us what is the probability that a healthy person becomes ill, in a certain period of time. Typically, K is the result of two factors $K = s\,p$, where s is the percentage of people a healthy person comes into contact with during that period, and p is the probability of person-to-person transmission of the infection. As it is easy to understand, the larger K, the faster the epidemic will spread. The passage from the infected compartment to the removed one also occurs with a certain probability r. The three interesting quantities in order to predict the future evolution of an epidemic are represented by the number of people present in each of the three compartments at a given instant of time: S, I, R. Their sum, of course, is equal to all the existing population since, each of us surely belongs to one of the compartments. The mathematical model consists of three equations that link these three quantities.

In the media these days a bell curve is often shown very similar to those I have drawn in Figure 2 and which represent the trend of magnitude I (number of infected) over time, in three different cases. I obtained these curves by solving the equations of the SIR mathematical model on a computer.

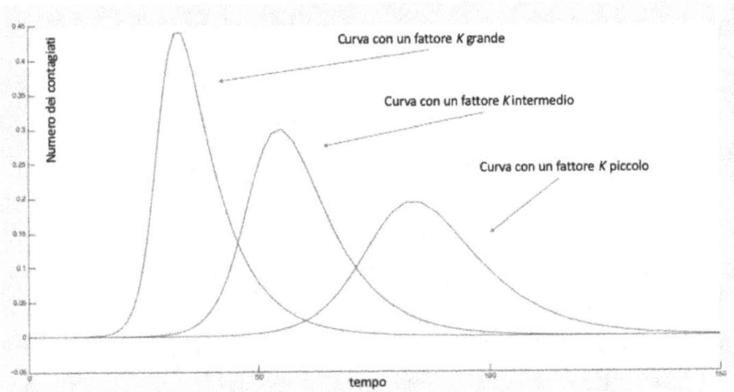

Figure 2 Number of infected versus time. Leftmost: curve with large K, center: medium K, rightmost: small K.

We can say that each of these curves consists of three parts: an initial part, characterized by a very fast growth in the number of infected people. Sometimes this part is called *exponential growth* , as it can be modeled with a simple exponential function. At the time of this article, most European countries are still in this phase of the epidemic. The central part, where there is the *peak of infections* and the final part, the *tail*, where the number of infected people begins to decline and then disappears completely. The specific shape of the curve and its amplitude depend directly on the value of those parameters that we have specified above, namely K (probability of getting sick) and r (probability of being removed from the sick, because they are cured and immune or because they have died). Although the SIR model that we have briefly illustrated is one of the simplest models, we can say that, even if with a certain degree of approximation, it still manages to grasp the essence of epidemiological models.

The three curves shown in the figure refer to three different values of the K factor. As can be seen, for smaller K values, two important benefits are obtained: 1) the reduction in the height of the epidemic peak, with a lower number of infected people in a given period, useful for not overloading health facilities. 2) The shift of the peak of the epidemic later in time, making the infection evolve more slowly and gaining precious weeks to prepare treatments and research vaccines.

How do you get a smaller K factor? It is necessary to act on those factors that make it up, therefore on s (percentage of human contacts in a given period) and on p (probability of transmission of the infection). As for s, European governments now seem almost unanimously oriented towards restricting the possibilities of movement (confinement at home) and closure of work activities. It is obvious that it is an effective strategy but that cannot be prolonged for long if not at the cost of serious restrictions on citizens' freedom and disastrous economic damage. As for p, however, we must orient ourselves towards the use of protective medical devices (masks, gloves, ...) and change of habits (no kisses and hugs and keeping distance).

As we have said, this SIR model is only one of the mathematical models in circulation and certainly the scientists who act as consultants for the health authorities have much more sophisticated ones. However, I hope it was useful to clarify how we approach an epidemic and what are the levers on which we can act to change its evolution, while trying to predict its future.

Appeared in Italian in N. 12/2020 *L'Osservatore magazine.*

5.   Excuse me, may I trace you?

"You are fighting a 2020 epidemic with the tools of the 1600s." It is one of the critical statements that occupy social networks these days.

In fact, the most used tools today to combat the Covid-19 epidemic, namely social distancing, the use of protective devices and quarantine for the infected, have an antique flavour. Already Alessandro Manzoni, speaking of someone infected with the plague of 1630, noted:

"Il tribunale della sanità fece segregare e sequestrare in casa la di lui famiglia; i suoi vestiti e il letto in cui era stato allo spedale, furon bruciati. Due serventi che l'avevano avuto in cura, e un buon frate che l'aveva assistito, caddero anch'essi ammalati in pochi giorni, tutt'e tre di peste. Il dubbio che in quel luogo s'era avuto, fin da principio, della natura del male, e le cautele usate in conseguenza, fecero sì che il contagio non vi si propagasse di più."

I promessi sposi, XXXI

(The health court had his family segregated and kept in his home; his clothes and the bed in which he had been in the hospital were burned. Two servants who had treated him, and a good friar who had assisted him, also fell ill in a few days, all three of the plague. The doubt that he had had in that place, from the beginning, of the nature of the evil, and the precautions used as a result, meant that the contagion did not spread any more.)

The effectiveness of these instruments depends crucially on the ability to apply them. In seventeenth-century Milan, it was not supposed to be large and in fact, despite these initial measures, the epidemic spread, causing over a million deaths, in northern Italy alone. Where are we now with their application?

Looking briefly at what is happening in the world, there are two distinct ways of dealing with the pandemic: the old *Manzonian* method of "close and segregate" and a new Korean method, vaguely inspired by Orwell's Big Brother, which we will call "trace and prevent". This second approach is also gaining consensus in our parts where the "close and segregate" has dominated so far.

What is it about?

The basic idea is as follows: since the infection is transmitted through close contacts between people, to limit the infection it is not necessary to segregate all people, it would be enough to segregate only the contagious people. The problem is that we do not know all contagious people and even when, through the execution of clinical tests, we discover one, we do not know how long it has been contagious and therefore how many people it may have infected up to that moment. For the quarantine mechanism to be effective it is necessary to have a method to be able to trace all the people who have come into contact with him or her. This task, which until now has been carried out simply by interviewing infectious people, could be made much simpler, faster and more precise using modern technologies based on the use of electronic movement tracking devices. Starting with our smartphones, to continue with our smart-watches, our laptops, our cars, all these devices are equipped with a satellite receiver for the GPS (Global Positioning System) signal that allows you to know their position in each point of the globe. A data that, if stored at regular intervals, allows you to reconstruct all the movements of the device.

The "track and prevent" strategy, to believe the results of the Korean data, would seem to work. So why not apply it also in Europe on a large scale? As you can imagine, there are a whole series of problems to consider, some of a technological nature, others more of a legal and political nature. The first technological problem is linked to the fact that not all smartphones use the same software and the same communication methods.

A step ahead in this direction has been made by Apple and Google: the two biggest players in the sector recently announced an agreement to make a tracking platform on Bluetooth technology operational, overcoming the differences between operating systems. Another technological problem is that not everyone has a smartphone. Especially in elderly people and children it is necessary to think about the use of other devices, such as electronic bracelets or "beacon" buttons to sew on clothes, preferably equipped with "energy harvesting" capabilities that make them energetically autonomous, without the need for rechargeable batteries. Some companies in the world of IoT (Internet of Things), such as the Swiss Aartesys and the Italian companies Vetrya and Wisepower (the latter is a spin-off company of the University of Perugia), have been working towards this goal for some time.

Then there are the legal and political problems: who has knowledge, control and ownership of the data relating to the movement of people? Who manages the database of infected people? You understand that detailed tracking of our movements constitutes a serious violation of our right to privacy. It is in this context, where law and technology intertwine, that some of the solutions available on the market today are measured. Some of these are directly borrowed from counter-terrorism practices and carry all of the typical characteristics. For example, the mandatory bracelets for the quarantine in force in Hong Kong (Euronews) or the platform made available by the Italian Cy4Gate of the Electronics Group, traditionally engaged in advanced technologies for military use and national security (Askanews).

There is no shortage of initiatives in the institutional sector as well. In Italy, the Ministry for Innovation has issued a call for technologies based on active monitoring of the risk of contagion. In a few days, 319 proposals arrived from companies, universities and research centers. According to Minister Paola Pisano, we will soon have an app to download on smartphones validated by the ministry. In Switzerland, too, attention is paid to the tracking problem. Swisscom reports gatherings of 20 or more people in public places to the federal authorities using cell phone data. The data is provided 24 hours late to protect privacy (Swissinfo.ch). At a European level, the Pan-

European Privacy Preserving Proximity Tracing (PEPP-PT) initiative is attracting the support of important research institutions such as the ISI of Turin, the German Fraunhofer and the Zurich Polytechnic. The project, called DP-3T, currently being tested is based on an application that uses locally encrypted data, without the need for centralized collection.

Indeed: encryption, on-device storage, and short-range communications (such as Bluetooth) appear to be three key ingredients for a successful technology in this area.

Finally, in order to make the "trace and prevent" approach truly effective, a wide diffusion of these devices is required, that is, the mass adoption by citizens. Such diffusion, in Western countries characterized by a strong awareness of rights, can only take place through voluntary acceptance, if some fundamental freedoms are not to be jeopardized. Let's not forget: there are rights (privacy, movement, worship, ...) that are no less important than the right to health. We cannot protect this by throwing those others away.

Appeared in Italian in N. 16/2020 *L'Osservatore magazine.*

6.   Who's afraid of 5G

5G, or the technology for cellular telecommunications of the future generation, which should begin shortly, is scary for many.

And I'm not just talking about the usual conspiracy theorists, flat-earthers & co., who see 5G as responsible for the Coronavirus pandemic and all the worst calamities. No, even serious and respected institutions such as the *Swiss Federal Office for the Environment* (FOEN) and the Italian CODACONS, have raised doubts and reservations which, in fact, have caused a partial moratorium on experimentation, both in Italy and in Switzerland.

What is people afraid of? In general, there are concerns about harmful effects on health, caused by the electromagnetic waves emitted by the antennas for 5G mobile telephony.

Rather than subscribing to the fans of those who are pro or con 5G technology, we would like to understand what the real substance of the debate is.

5G, as the name implies, is a new technology that should replace 4G, which is still in use today. Which in turn replaced 3G in 2012, which replaced 2G in 2001 which since 1992 was the successor of the first cellular communication technology NMT (Nordisk MobilTelefoni) activated in 1982. The reason why every few years a new technology replaces the previous one, it is easy to say: we want to have telecommunications that allow us to transfer a greater quantity of information, in the shortest possible time, at the lowest possible cost. Compared to 4G, 5G let us connect many more devices to the network at the same time (something that will allow us to give life to the so-called "Internet of Things"), go at speeds at least ten times faster in data transfer and consume a lot less energy. Several applications of the near future, such as self-driving cars, personalized healthcare, electronic money, artificial intelligence, depend on the advent of this new technology.

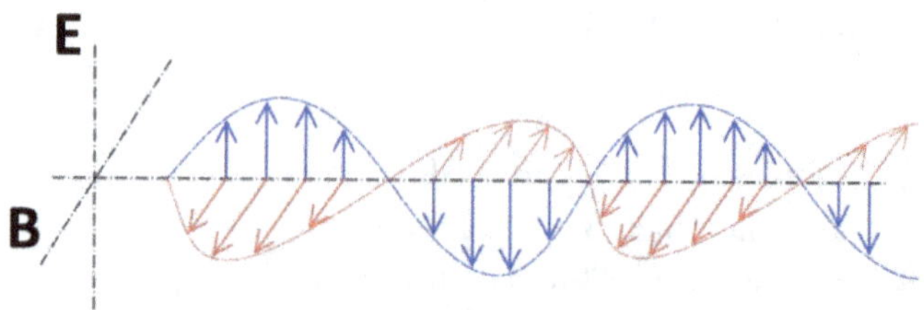

Figure 1. Representation of a plane electromagnetic wave, with its sinusoidal components of the electric field **E** and magnetic field **B**.

Let's come to health concerns.

In fact, we know quite well the effect that electromagnetic waves have on biological beings, because we have been studying them for at least 120 years, since the German physicist Rudolf Heinrich Hertz, at the University of Berlin, provided proof of their existence. We need to keep an eye on two important parameters: one is the frequency of the wave, the other is its intensity. With reference to Figure 1, an electromagnetic wave, represented here as a plane wave, is actually the composition of two sinusoidal functions, one for the electric field **E** and one for the magnetic field **B** (hence the name of Electro-Magnetic wave).

The wave frequency indicates how often the sinusoid crosses the horizontal axis in a given time interval. The more often it crosses it, the higher the frequency. In figure 2 you can see the sinusoid superimposed on a colored panel. It represents light (which is also an electromagnetic wave): the red color is a wave with a lower frequency than the purple color.

The intensity of a wave, on the other hand, is linked to the amplitude of the sine wave. If I have two waves, one with amplitude 1 and the other with amplitude 2, then the second has an intensity four times greater than the first. One thing to keep in mind is that the intensity varies with the distance from the source of the wave. If I am one meter away from the antenna I have a certain intensity; if I move away from ten meters, the intensity generally decreases by a hundred times. I wrote "generally" because, as we will see in a few lines, this does not always happen in 5G.

So, in summary, to understand if an electromagnetic wave is harmful to health we need to know its frequency and its intensity. With respect to frequency, the waves are generally divided into two classes: ionizing and non-ionizing. The dangerous ones are the ionizing waves, because, one they hit biological tissues, they can tear electrons from the atoms that make that up. They typically have higher frequency than ultraviolet rays, such as the fearsome x-rays and gamma rays.

5G waves, on the other hand, belong to the class of non-ionizing ones and are considered not "immediately dangerous" for living organisms. At this point, however, intensity comes into play, because even non-ionizing waves, if sufficiently intense, can cause damage. As know well those who fall asleep unexpectedly under the sun on an August day: high intensity and long exposure time can leave their unpleasant mark.

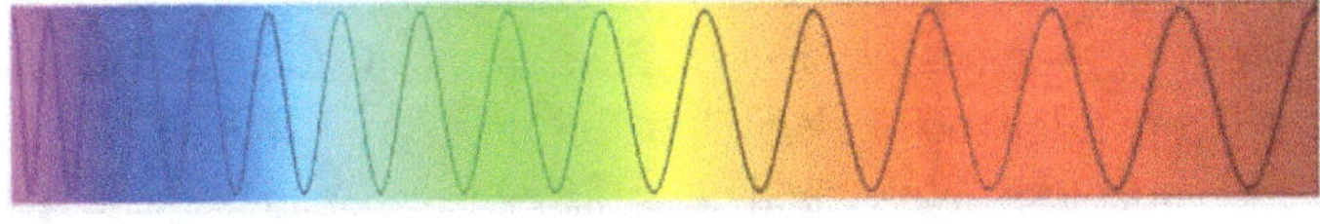

Figure 2. Light is an electromagnetic wave made up of many frequencies. The red colour has a lower frequency than the purple colour.

5G employs waves with an intensity even lower than that of 4G. So, in principle, there should be no cause for concern. However, some observations must be made that can help us better understand the point of view of those who oppose 5G.

The first note concerns the antennas. While a 4G antenna emits its waves simultaneously in (almost) all directions, a 5G antenna has the ability to focus this emission towards a single mobile phone. Thanks to this ability (called beam-forming), it can emit waves at a lower intensity because it avoids wasting energy in directions where there is no receiver. However, the fact that the waves are "focused" can rise the suspect that their intensity might be high and provoke the request for "further investigations" made by some consumer associations.

The second note is about the actual danger of non-ionizing waves. The controversial (according to US) *World Health Organization*, through its *International Agency for Research on Cancer* (IARC), has classified radio frequency waves (4G and 5G) among the "possible human carcinogens" on the basis of studies not yet completely conclusive. Instead, other agencies such as the US EPA and the powerful *Food and Drug Administration* (FDA) have chosen not to consider them carcinogenic.

For my part, I can tell you that years ago, following studies conducted on a phenomenon known as "Stochastic Resonance", which showed how, in the presence of non-linearity, even a small periodic signal was potentially responsible for large effects on the dynamics of a physical system, I was contacted by groups of self-styled scholars. They wanted to convince me to apply my studies to the effect, according to them too neglected, of low intensity waves on biological systems. Needless to tell you that I refused because their motivations, far from being completely disinterested, were also quite weak from a scientific point of view.

Years later, I haven't changed my mind.

Appeared in Italian in N. 21/2020 *L'Osservatore magazine*.

7. Post Covid-19: the world to come

In these days when, at least in Europe, the epidemic data make us breathe a sigh of relief, the debate on what the post Covid-19 world will be like is growing.

Let's try to make some (risky) predictions:

A) Scientific knowledge will play an (even) greater role.

Last week I was in a public debate which Michele Mezza, journalist and professor of Marketing and New media at the Department of Social Sciences of the Federico II University of Naples. He did declare that this is the first epidemic in history that has been cured by physicists and mathematicians, rather than by virologists.

Although I believe that this is an exaggeration, it is interesting to note how, in public opinion, there has in fact been a change of perspective on the role of science, with particular reference to the so-called "hard sciences" (mathematics, physics, chemistry, biology ...), which has important consequences for the world to come, with all due respect to the conspiracy theorists and flat-earthers (except for the Pastafarians because they have style in the choice of headgears).

As a matter of fact, we too must note that, now three months after the piece we published an essay in issue 12 of *L'Osservatore Magazine* (see *4. Predicting the future of the epidemic*), the mathematical models of the evolution of the contagion, have proved to be largely correct and of great utility in guiding the choices of governments. Whatever the virologist Guido Silvestri says. He, interviewed by La Stampa on June 8 2020, invited us to take note of the failure of mathematical models and to which, rightly, the Italian Mathematical Union replied, two days later, with a dry statement: " A mathematical model is not a crystal ball. It is a tool that allows us to objectively calculate the consequences of what we know about the transmission of the virus ». I subscribe.

B) The geography of living: from the centre to the periphery.

In times of reduced mobility, there are many who have appreciated the possibility of not having to physically move. To this must be added the undeniable advantage of living outside the big cities and/or in the presence of large green spaces. These are experiences that leave their mark in the long term and will deposit some important values in the memory of the youngest. We will see the consequences in the decades to come, but it is easy to predict that the process of moving from the city center to the outskirts, already widely present, will accelerate. On this theme, the authoritative and passionate intervention of architects such as Stefano Boeri and Massimiliano Fuksas, who call for a return to "small villages", should be noted. These are locations that are certainly not lacking in a large part of Italy and Switzerland.

On the other hand, it is necessary to ensure that the geographical distance is not also social isolation, ending up favoring those who already have networks of consolidated relationships over those who are entering the world of work today. Vincenzo Galasso rightly noted this point in the *Plusvalore* broadcast of 5 June, on the second channel of CSR. To avoid these risks, it is important that no territory remains isolated.

Anna Laura Orrico, undersecretary of Mibact with responsibility for relations with Regions and local authorities, interviewed in Repubblica by Cristina Nadotti, proposes *a relaunch plan to promote a decentralized system, essential for a more sustainable lifestyle. Especially on digitization, the plan to bring ultra-broadband to 6500 small municipalities has already been launched, the works have already advanced in 2500 of these and will be completed everywhere by 2023. The State is taking charge of covering the so-called "white areas" those that private companies do not cover because they are not profitable .*

C) The environment as a common good.

Whatever the link between the spread of the epidemic, pollution and environmental conditions in general, there is no doubt that the desire to live on a cleaner and healthier planet has grown in public opinion in this period. A theme close to the heart of this pontificate, as the President of the Pontifical Academy of Social Sciences, Stefano Zamagni authoritatively recalled, last June 1st, during the conference held online for the Osservatore Democratico of Massagno, as reported by Markus Krienke, in issue 23 of this magazine: *In this way, we move from the imperative of maximizing the "total good" to the possibility of realizing the "common good" which, through respect for the limited nature of the environment, is also in solidarity with future generations.*

These are important and demanding words.

For example: during the period of closure dictated by the pandemic it was observed that «one of the most visible consequences of the crisis was the decrease in energy needs and the collapse of oil prices. However, when economies recover, CO2 emissions are likely to return to past levels and even surpass them, jeopardizing the transition to a low-carbon economy "(Virginie Fauvel, Chief Transformation Officer of Euler Hermes) .

In order for the emotional thrust not to be thwarted, it is important that the economic revitalization programs are strongly inspired by environmental sustainability and the common good.

Obviously, there are others and much more solid forecasts circulating in the national and international media these days, and not all of them have this sweet and optimistic character. Here we tried to make a game that I invite readers to continue, alone or in company.

How Alan Kay, one of the pioneers of the digital revolution in the seventies (remembered by Federico Ferrazza on Wired last

December) does keep reminding us: The best way to predict the future is to invent it.

This awaits us: let's roll up our sleeves.

Appeared in Italian in N. 25/2020 *L'Osservatore magazine*.

8. How intelligent is Artificial Intelligence?

I read that Jared Kramer, Amazon's Machine Learning Manager III (whatever it means) said they are testing a new chatbot to manage customer relationships (Ben Stevens on *Charged*, Feb. 27, 2020) and I'm suspicious: could it be that Miss Mirela, who handled my request on the late order of two fruit juicers, instead of being an Eastern European lady with uncertain syntax, is instead an artificial person, that is an artificial intelligence (AI) algorithm programmed to chat with me with undisguised condescension?

I let you judge for yourself. Here is the faithful transcript of my conversation:

Initial question: *Hi, I ordered a product (2 juicers) on June 2nd and it was supposed to arrive on June 9th but it didn't arrive. The courier website (Deutsche Post DHL, tracking CS2070016275) says: Waiting for pickup - 04-06-2020 15:11.*

05:16 PM Mirela (Amazon): *Good evening, thank you for contacting Amazon.co.uk customer support. I'm Mirela and I will be happy to assist you. Who do I have the pleasure of chatting with?*

05:16 PM Luca Gammaitoni: *Luca Gammaitoni*

05:17 PM Mirela: *Before offering you my assistance, I would like to anticipate that if the chat is interrupted, I will immediately contact you by phone on the number in your account to make sure that you do not have to repeat everything again.*

05:17 PM Luca Gammaitoni: *ok*

05:18 PM Mirela: *Hi Luca, can you confirm that it is this order? Panasonic MJ-L501WXE Slow Juicer, Bladeless Juicer, Slim and Elegant Space Saving Design, Extremely Quiet, Soft Finish, 150W*

05:18 PM Luca Gammaitoni: yes. *I ordered 2 of them*

05:19 PM Mirela: *I would kindly ask you to give me 2 minutes so that I can carry out a further verification*

05:19 PM Luca Gammaitoni: *ok*

05:21 PM Mirela: *I'm sorry for this inconvenience with your products, the packages have been lost by the courier. I can proceed directly with the refund*

05:21 PM Luca Gammaitoni: *ok*

05:22 PM Mirela: *Where do you prefer to be released the amount via Amazon gift card or card as the original payment method?*

05:23 PM Luca Gammaitoni: *card*

05:24 PM Mirela: *It will take up to 5-7 business days for your refund to be visible*

05:25 PM Mirela: *Can I do anything else for you?*

05:26 PM Luca Gammaitoni: *no, thanks.*

05:27 PM Mirela: *It was a pleasure, good evening!*

05:27 PM Luca Gammaitoni: *good evening*

Whatever you think, the idea of having unknowingly conversed with a computer program seems a little disturbing to me. And think that I should be used to it: the first time I wrote about "artificial people" was in March 1997, when *Avvenire* (Italian national newspaper) had the goodness to publish a piece of mine on the first steps of what would become an AI flourishing trend.

In the meantime, many things have changed; these programs that interpret natural language have become much more sophisticated and today we call them *chatbots*: *chat* in the sense of chat and *bot* in the sense of robot.

Who knows what the android Bishop (actor Lance Henriksen) would think of this nickname. He who, in *Aliens*, the second film of the successful series, had angrily replied to Lieutenant Ripley (Sigourney Weaver): *I prefer to be called an artificial person.*

Be that as it may, today we have chatbots and among the most popular are certainly Apple's Siri, Amazon's Alexa and the Google's Meena (three female names: a coincidence?) which promise performance even closer to human dialogue.

How do these amazing artificial conversationalists work? They all use a technology known as *machine learning* (or, in its most recent version, *deep learning*). To explain what is this technology, let's take for example one of its most popular applications: image recognition.

Suppose we want to make a program that is able to recognize whether the digital image we have contains a cat or not. This problem is relatively simpler than that of natural language recognition, but the principle of operation is the same.

To achieve this, we start by writing a program that contains a definition (= model) of a cat (example: *an object with 4 legs, two ears, whiskers, tail, ...*). Then, a large number of photographs with a cat are provided to the program and asked to locate the cat. When the program fails, it is told where to change its model. This procedure is repeated many thousands of times until the model is refined to the point that it now gets it right more than nine times out of ten. The idea behind this approach is to imitate the learning of a child who learns to recognize a cat using examples provided by experience. Obviously, this program will work very well in the end, even if only for the identification of cats. In this sense we can say that this very powerful technology is nothing more than a *classification scheme*, albeit very efficient.

In recent times it has begun to be debated whether this *classification scheme* can be used to generate new knowledge, as would a truly intelligent being. A concrete example is the discovery of new laws of nature. If we provide a lot of experimental data to an AI program that uses deep learning technology, is it able to discover a law that we don't yet know?

This question, anything but trivial, has recently been the subject of great debate between the supporters of the yes (Big Data approach) and the supporters of the no (mathematical model approach).

The debate is still very open and there are truly surprising reflections to be made, which point right to the very heart of knowledge and of what is knowable. It is a really interesting story and, if you have the patience to wait for the next episode, you will discover it with us.

Stay tuned.

Appeared in Italian in N. 29/2020 *L'Osservatore magazine.*

9. As to why it's impossible to find something if you don't know what you're looking for

Cuando se proclamó que la Biblioteca abarcaba todos los libros, la primera impresión fue de extravagante felicidad. Todos los hombres se sintieron señores de un tesoro intacto y secreto. No había problema personal o mundial cuya elocuente solución no existiera: en algún hexágono.

J.L. Borges, La biblioteca de Babel

Last month, in the wake of the reflections on the Artificial Intelligence (AI) applications we deal with every day, from chat bots (automatic verbal responders) to medical diagnosis systems, to automatic guided systems, we had set ourselves a question: how intelligent are today's AI applications, based on *deep-learning*?

More specifically, we want to address the task of creating new knowledge, like discovering a new law of nature. As a matter of fact, creating new knowledge or inventing/discovering new things is considered an exclusive human activity. How can AI perform in this kind of tasks?

I recently talked about this topic at dinner with a colleague and friend, Barry Barish, Nobel Prize in Physics 2017.

He put forward the hypothesis that deep-learning could be used to discover new forms of gravitational waves, still unknown. Since gravitational wave interferometers make continuous observations in search of any signals, they end up accumulating large amounts of data that are continuously analyzed in search of known waveforms, as the waveform associated with the collision of two black holes or with the explosion of a supernova.

Left to right: Luca Gammaitoni, Barry Barish. Perugia 2018

What if, hidden in that data, there were waveforms relating to phenomena that we do not yet know? How can we go about identifying them?

The idea, per se, is not new.

Starting with the article by Chris Anderson, editor-in-chief of Wired, who in 2008 wrote a piece entitled *The end of the theory: the deluge of data makes the scientific method obsolete*, the belief has grown among professionals that the great availability of "experimental" data allows us to overcome the traditional work of scientists who, to generate new knowledge, build mathematical models by mixing observation, intuition and creativity.

For example: if we provide a lot of experimental data to an AI system, which uses deep-learning technology, will it be able to discover a law of nature that we do not yet know? Could it discover a cosmological event that generates gravitational waves and which is unknown to scholars? This would be a beautiful display of intelligence, which would place an artificial system in direct competition with human intelligence.

Unfortunately, using deep-learning technology which, as mentioned, is a sophisticated classification scheme and nothing more, it is not possible to produce new knowledge, in the sense just described. The reason why this is the case, is very interesting and points to the heart of what links the *meaning* (metadata) of an expression to its *formal representation* (data).

To understand, let's start with an example. Let's consider a string of numeric symbols: 31415926535. These symbols constitute the starting data, they are no different from those that are provided to AI programs every day, except that here there are only twelve digits while usually billions or trillions of them are analyzed at once. What does this data represent? Difficult to say, they could be the quantity expressed in Francs of the Swisse cantonal budget, or they could constitute an advertising brand (*invicta 31415 Men's 63mm Reserve Grand Arsenal Swiss Chronograph*), they could be the encryption of a message (*Cada ei B fece*; using a simple code based on substitution, the message takes on a vague political meaning), or, the first twelve digits of the number pi, represented in base ten. How do you know what the right interpretation is?

Answer: you can't.

The very reason why this is not possible, is well explained in a booklet written a few years ago by a great Italian mathematician, Lucio Lombardo Radice, entitled *The Infinite*. Lombardo Radice analyzes several interesting examples. The most intriguing of all is the one illustrated in the famous story by Jorge Luis Borges, The Babel's library.

The books in the library of Babel consist of four hundred and ten pages; each page, of forty lines; each line, of forty letters. Their content consists of sequences of characters written at random: the library contains all possible sequences, each in a different book. In this way the library contains everything that can be written in a book, in any book of 410 pages. If there is everything, then there will also be an answer to the questions of each of us.

Borges writes:
When it was proclaimed that the Library included all the books, the first impression was one of extraordinary happiness. All men felt themselves the masters of an intact and secret treasure. There was no personal or global problem whose eloquent solution did not exist: in some hexagon.

To our great disappointment, however, the situation is slightly more complicated.

Lombardo Radice reveals Borges' vertigo deception: even if the number of books in the Library of Babel is very large but still finite, the number of meanings that can be attributed to the text of those books is infinite and therefore cannot be contained in those books.

How is it possible? It would seem a paradox. And indeed, it is: it is called Richard's paradox, from the name of the French mathematician Jules Antoine Richard (1862-1956) who showed how a finite string of characters potentially corresponds to an infinite number of meanings. In other words, having a certain set of data available, is equivalent to having the potential answer to an infinite number of physical problems. In other words: for every possible answer there is an infinite number of perfectly compatible questions. What is the question that you are interested in? It is impossible to know the answer if the question is unknown to you. There is no algorithm or computing power that holds, however large it may be, it will not be able to find the answer to an unknown question.

That's why it's impossible to find something if you don't know what you're looking for.

Does this mean that no AI will ever be able to equal man in the search for knowledge? No.

This just means that the present generation of AI programs, based on deep-learning technology, is far from succeeding.

How about the future? ... stay tuned.

Appeared in Italian in N. 33/2020 *L'Osservatore magazine.*

10. Is man a machine?

Then the Lord God formed a man from the dust of the ground and breathed into his nostrils the breath of life, and the man became a living being.

(Genesis, 2,7)

In this essay we will keep dealing with Artificial Intelligence (AI). Let's try to go beyond *deep learning* algorithms, of which we have seen the inevitable limitations in the previous chapters.

If, as acquired, the meaning cannot be uniquely associated with a string of characters (or a series of words), if only the data and not the meanings can be stored in digital memory, so how is meaning generated? How is it possible that we human beings are able to create new meanings and associate them with strings of characters? And, above all, is this ability only ours or could it also be the prerogative of artificial intelligence systems?

We will try to answer these questions in this chapter.

Let's start with the first question: how do you create the meaning of a character string? Let's take the Italian word "scudo" (plural: scudi) for example. "Scudo" in Italian is used to indicate a *shield* (as used by the ancient Greeks in a battle) and, at the same time, is the name of a popular renaissance coin. The phrase *I have ten scudi with me*, would have had two very different meanings if pronounced during a meeting of the Achaeans at the time of Homer or in a street in Bologna in the early 1800s. How can we interpret it unambiguously?

Simple: we can't.
To "give meaning" to the string of characters we need, in fact, an entire world of reference, a real world made up of relationships, things, feelings, memory ... The creation of meaning is a process that

therefore cannot be contained in the simple sequence of characters that make up a sentence, however long the sentence is. Ergo, as we have discussed, thanks to Richard's paradox, we cannot uniquely associate a meaning to a sentence. The meanings are enormously superabundant (infinity of higher cardinality) than the sentences and, above all, they belong to a real world, with all the complexity connected to this condition.

Having ascertained this, let's move on to the second question, certainly more demanding: is this ability to generate meanings only of human beings or can it also be the prerogative of artificial intelligence (AI) systems?

About this point, it must be said, opinions are very diversified and, entering into this debate is equivalent to walking a kind of path inside a minefield. However, I do not want to escape the question and I will tell you what I think: I am definitely against statements such as *machines will never be able to equal man in the production of thoughts, feelings, creation of new meanings.*

I think that, probably, those who are lucky enough to live long enough will see the day (still distant) in which we will have AI systems capable of creating meaning, that is, of possessing the same abilities that we attribute to men today.

We will have the poetic cyborgs like *Blade Runner* and the assassin ones like *Terminator*, the sensual and self-aware androids like Ava from *Ex Machina*. In short, we will have the whole package of what science fiction has been able to imagine (including risks).

Why do I think this? The answer is simple: since man is a machine, sooner or later we will be able to build machines similar to man and therefore with his same capabilities.

I'm not naive and I understand that the statement "man is a machine" is capable of raising an eyebrow in more than one of our readers. However, it is good that you know that the vast majority of scientists (and I am one of them) today share this statement.

It is another thing to ask whether man is *only* a machine or is there something else. Catholics (and I am one of them) believe that there is more. What more? There is that man, beyond his nature as a machine, or a wonderfully complicated assembly of atoms, also has a *soul* capable of distinguishing him from other animals and, more generally, from any other living being. Unfortunately, the discussion here becomes fraught with difficulties, because any sensible reflection on the subject of the *soul* would require much more space than is available here and skills that I certainly do not possess.

I would just like to point out that there are two meanings currently present in traditional Catholic doctrine. First, soul as the engine of life (it is no coincidence that a living being is called animated from the Latin *anima*, i.e. *soul*). Second, soul as a sign of the Spirit, or what makes us similar to God (When God created man, he did it in the likeness of God, Genesis 5,1).

The first meaning, clearly of Aristotelian matrix, has today been excluded from modern science which has shown how life is not the fruit of some unknown ingredient but is the manifestation of microscopic molecular machines that combine to give motor to all the processes necessary for growth and reproduction. Molecular machines are aggregates of atoms that, at the nanometer scale (there are about ten atoms in a nanometer), move to perform simple tasks, animated by thermal noise, or by the incessant fluctuation that characterizes all microscopic things.

In this sense, life is nothing more than the manifestation of the dynamics of a physical system (generally rather complicated) that is found to operate out of thermodynamic equilibrium, driven by the omnipresent random forces (noise) that characterize all matter. If this definition does not seem poetic to you, I do not blame you. To me, to tell the truth, it looks beautiful and makes me shiver every time I write or say it, but I understand that there can be different sensitivities.

Be that as it may, in this sense, the soul has nothing to do with life.

Thus, we are left with the second meaning of the term *soul*.

As the Bible narrates, the infusion of the soul into man is a gratuitous gesture of God. Therefore, it is not inherent to his being a machine (in fact animals, also machines, do not have it). Thus, I propose a fair distribution of tasks: it is up to us to understand the world and design new and increasingly intelligent machines, capable in a still distant future, of supporting man in his work of creating meaning. To God is to infuse the Spirit (to the machine-man or to another machine, of His choice).

In short: to each his own, without wanting to explain to God what He must do.

Appeared in Italian in N. 37/2020 *L'Osservatore magazine*.

11. Black hole is not a hole

In this episode I would have gladly continued to explore the aspects connected with the "mechanistic" interpretation of life to convince you, if anything were needed, that man is actually a machine, even if it is a truly extraordinary machine. Instead, the austere and unfathomable gentlemen who populate the Royal Swedish Academy of Sciences awarded the Nobel Prize in Physics 2020 to the subject of black holes and, for once, I decided to succumb to the indiscreet charm of current events.

I confess that at the announcement of the names of this year's Nobel Prize winners in Physics, I was a little surprised. In fact, in the last four years they have awarded the prize three times to the same branch, that is to astrophysics: in 2017 gravitational waves, in 2019 it was the turn of exoplanets and this year of black holes. It is quite unusual for the Royal Academy to linger for so long in the same area of Physics.

I owe my limited knowledge of black holes to Kip Thorne (Nobel Prize 2017) who, during a crowded public conference in Pasadena, many years ago, holding a small rubber ball that he had specially painted black, said: *everyone thinks a black hole is a very mysterious thing but, instead from a close inspection, it looks a lot like this ball.*

This has remained in my mind ever since: a black hole is not a hole, it is a ball of very, very dense matter. So dense that its gravitational field is so strong that it significantly warps space and affects time.

This *curved space* thing is Einstein's idea. Gravity, which everyone from the time of Newton onwards, thought was a force capable of attracting mass, is not a force, it is only a manifestation of the curvature of space. To understand it better: if we throw a ball on the surface of a table, it will travel in a straight line. If, on the other hand, we throw it on the surface of a funnel, it will tend to curve down towards the centre of the funnel itself. The reason why it curved is that the surface of the funnel is inclined and we don't need further explanations to figure it out. The same thing is true for gravity: the presence of mass curves space and whenever we are close to a large mass, such as a planet or a star, this curvature is felt and all objects tend to move along lines that hold account of this curvature. Since it is the space that bends, then also the light that is known to move in a straight line in the flat space, is forced to move according to the curve it assumes.

This certainly curious idea of curved space is part of the work that Einstein published in 1915, together with the famous equations that rule such a phenomenon. These equations are, from a mathematical point of view, rather complicated to solve and Einstein himself, who was not a great mathematician, was unable to find an exact solution. The first to find it, was a fellow countryman, Karl Schwarzschild, who published it in the same year, while he was at war on the Russian front. Schwarzschild died the following year, in 1916, but his work soon became famous. One of the consequences of this work is that, given a certain mass, if this is very dense, then there is a distance (which today we call the Schwarzschild radius) within which space is *so curved* that nothing can escape, not even light. The Schwarzschild Radius defines what is called the *event horizon* of a given body with mass.

For most material bodies, characterized by not very high densities, this radius is very small and has no practical effect. For black holes, on the other hand, as these are extremely dense bodies, the Schwarzschild Radius is located at a certain distance outside the body and the event horizon becomes a region surrounding the body, producing the remarkable effect of light trapping we have been talking about.

The need to find easier mathematical methods to explore this story of the event horizon leads us to this year's Nobel Prize, awarded to the already very popular British scientist Roger Penrose. In 1965

Penrose proposed a new method for dealing with Einstein's equations and demonstrated that there are solutions that predict the existence of objects so dense that are capable of showing the effects predicted by Schwarzschild. These objects, thanks to the American astronomer John Wheeler who in 1967 invented the name, are called *black holes*.

Penrose has gained enormous media popularity for many years, thanks also to its intense and valuable activity of popularizing science. I myself witnessed it. In December 1997, I was in Pune, India, participating at an international conference (International Conference on General Relativity and Gravitation GR15) devoted to all topics associated with gravity. At that time, I was involved in the research of gravitational waves and with my group we were studying a method to increase the sensitivity of the Italian-French detector called Virgo, by limiting the effects of thermal noise on the optics of the interferometer.

The international conference included a public lecture by Roger Penrose, to be held in an auditorium in the city. The organizers had set up buses to transfer conference participants to the auditorium. When we were near our destination, the bus was forced to stop and we had to walk, escorted by the police, the last five hundred meters, because a large crowd of people had formed around the auditorium who wanted to enter to listen to the lecture by Penrose. Having managed to enter the auditorium we could not find seats because they were all already occupied and we were forced to do as best as possible on the steps of the access staircase. As Penrose began the conference, outside the police tried to calm the several hundred people who had not been able to enter and who were hotly protesting.

I have never seen such a manifestation of popularity by a scientist and, alas, I fear, it will never happen to me.

Pubblicato nel N. 37/2020 de *L'Osservatore magazine*.

12. Computers as small as specks of dust

To the audience of students who look at me curiously, the enlarged image of the fingertip arouses surprise and a little hilarity. Even more surprising is the question I ask them: *what do you see in this photo?*

After some hesitation, a girl in the third row raises her hand and says:

a finger with two dots, one dark and one light.

Right. What do you think these dots are? I pursue.

Silence. After a moment of general embarrassment, I resume: *the clear one is a speck of dust. The darker one... a computer.*

Surprise, disbelief, admiration.

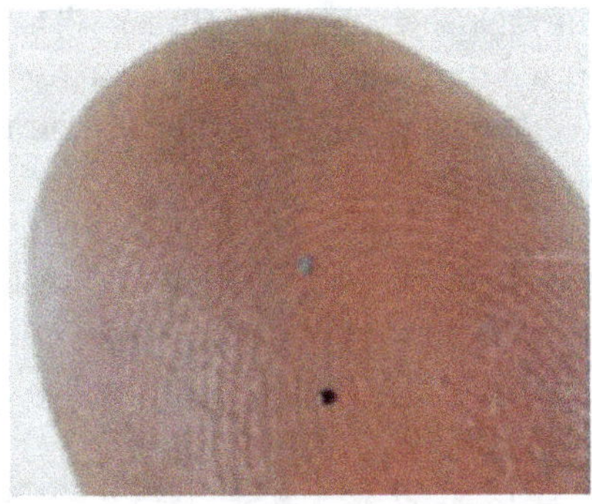

Yes, for some time now humanity has had the technology to build computers that are as small as specks of dust. A technology that we are using to build very small machines capable of measuring physical quantities and communicating the result of these measurements. We call them *wireless sensors* and have dimensions ranging from a few centimeters to a few tens of microns (thousandths of a millimeter), just as large as grains of dust.

Their purpose will be to help us live better. How? For example by monitoring our state of health, or the stability of bridges and buildings, or by checking our home, cars and appliances. As an example of application, we could use them inside clothes, inserted in fabrics, ready to regulate the passage of air according to the external temperature and the degree of sweating of our body, increasing or decreasing the distance between the fibers. We could paint them on the walls of our home to check their stability and the possible presence of moisture on the plaster. We could spread them on our skin to measure the temperature of the body or the degree of dryness of the skin or pass through the hair to change its colour. Not to mention biomedical applications, where wireless micro sensors could soon be inserted inside the human body, for targeted release of drugs or as micro-robots capable of carrying out specific operations.

This is far from being a mere dream. The market for these sensors is expected to reach $ 148.67 billion over the next six years with a CAGR (compound annual growth rate of investment) of 18.3% (Fortune Business Insights, April 2020).

How is it possible that computers of such small size can be built? We owe this exceptional result to the research that has developed, mainly in the USA, after the end of the Second World War onwards. The starting date of this adventure is 1947. In that year three American physicists: Walter Brattain, John Bardeen and their leader William Shockley, invented the *transistor*. A device obtained by suitably contaminating Silicon or Germanium crystals with Boron or Phosphorus. For such an invention they would get the Nobel Prize in Physics in 1956.

Since its invention, the transistor has become the building block for all computers. The story of the development of such a device is very interesting. Between the end of the 1940s and the end of the 1950s, started a race to see who could make smaller transistors. The smaller the more you could fit in a given space. Unfortunately, as you may imagine, there are limits to what can be done by welding the individual electronic components by hand in order to make complex circuits. In 1958 two people, apparently independently, came with an idea that would have changed the future of the transistor and ours as well. Instead of building single transistors and then welding them together with conductor wires, why not try to make everything on a single silicon plate, using metallic liquid deposited on the surface of the plate instead of the wires? It is the birth of the integrated circuit. The two scientists that came up with such a smart idea are Jack Kilby, then recently hired at Texas Instruments and Robert Noyce who worked at the start-up Fairchild Semiconductor.

Starting from the 1960s, the race for pushed miniaturization (small, large and very large-scale integration) begins. The goal is to increase the number of transistors per square millimeter from a few units up to one hundred in the early 70s, ten thousand in the early 90s and gradually growing exponentially up to ten billion transistors in recent years. A growth that is sometimes referred to as Moore's Law, from the name of one of the founders of Intel, the company that first built an entire microprocessor (the heart of the computer) on a single silicon chip.

Increasing the number of transistors has obviously increased the computing capacity of a computer and decreased the size of these fundamental components. It has gone from single centimeter-sized transistors to a few millimeters in integrated circuits, down to the present 10 *nm* (*nm* = nanometer = billionth of a meter).

Can it be further decreased?
Scientists all over the world have been trying to answer this question for at least twenty years. Already six years ago, I remember seeing in Cambridge (UK) a prototype, made by Hitachi, of about 7 *nm*. It was an experimental device unsuitable for commercialization because it was not yet fully functional. The reason is that when we are

dealing with such a tiny transistor, the number of atoms composing it begins to become relatively small. A row of 20 silicon atoms in a crystal occupies about 7 *nm* and with such small numbers, the theory of operation of transistors, known as the *physics of semiconductors*, begins to no longer apply. For this reason, scientists are moving towards abandoning transistors to build binary switches of different types. Today we have several prototypes of new devices ranging from *nanomagnetic* switches (functioning according to a theory called *spintronics*) to *optoelectronic* ones, up to *electromechanical* ones made with carbon nanotubes.

For all these new prototypes, given their dimensions approaching those of a single atom (one tenth of a *nm*) it is necessary to use quantum mechanics, a theory that has the taste of something new and very difficult but is actually an old theory of more than hundred years, though still mysterious.

We will talk more about it, stay tuned.

Appeared in Italian in N. 46/2020 *L'Osservatore magazine.*

13. Waves of epidemic

Among the few things we know, analysing the epidemics of the past, is that these occurred in successive waves, spaced out over the time of a few months and that they lasted a total of a few years.

The current Covid-19 pandemic is no exception: in Europe we are recently in the waning phase of a second (and worse) wave and we are wondering if we should prepare for a third wave. It will be? How large and dangerous will it be?

In order to answer this question, we must go back to epidemic models, as we discussed previously (see L'Osservatore, No. 12/2020 and further back in this book).

The simplest model, however capable of capturing the salient features of an epidemic, as we have seen, is the SIR model. Here, the entire population of a certain homogeneous region (in our case at least one continent, given the possibility of people to travel and come into contact with each other) is supposed to be of N people, divided into three groups: those who can get sick, indicated with S (Susceptible), those who are currently ill, indicated with I (Infected) and those who are healed or died, indicated with R (Removed). At the beginning of the epidemic $S = N-1$, $I = 1$ and $R = 0$. As we saw previously, in the SIR model there are basically two parameters that matter, one is related to the probability of being infected (we had called it K) and one is related to the probability of being removed (because healed or died).

If these two parameters were constant over time then the epidemic would not show the phenomenon of waves. Simply, once started, this would go on in time until most of the population became infected and then slowly it would tend to disappear due to lack of people to infect. In this situation, if we look at the number of infected I, it would show only one large peak, so only one wave. However, the price to pay would be linked to the fact that a very large percentage of the population, close to 100%, would become infected. To be rigorous, in a more realistic model, it is not necessary that the percentage of susceptible is exactly zero but it must certainly be much lower than that of people already infected: $S << I + R$.

This scenario, sometimes called achieving *herd immunity*, would seem to have been taken into consideration during the early stages of the epidemic, in countries such as Great Britain and Sweden, that however soon shelved it due to the large number of predictable victims.

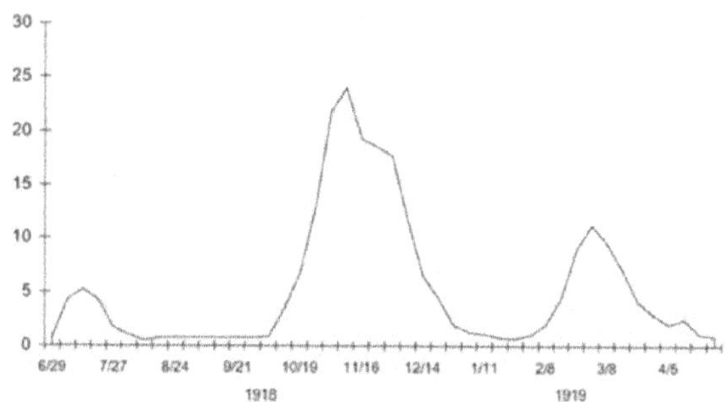

Figure 1: Three of the successive waves of the Spanish epidemic. In the vertical axis the number of deaths per 1000 inhabitants, in the horizontal axis the months. Da Taubenberger JK, Morens DM. 1918 Influenza: the Mother of All Pandemics. Emerg Infect Dis. 2006;12(1):15-22.

It is important to understand that, in any epidemics, the only reason why an epidemic can actually end is due to the fact that the number of S people, that is, the healthy ones who can be infected, must be greatly reduced. This is due to the fact that the development of an epidemic resembles the spread of fire in a bonfire: it starts slowly and grows rapidly, fuelled by the available wood, until it reaches the maximum, then it tends to decrease and go out as soon as it has consumed all the available wood not yet burned. Similarly, the epidemic goes out only by removing available firewood, i.e. healthy people to be infected. At this time in Europe, as in other parts of the world, despite the great spread of the disease, we are still very far from having reached this condition ($S << I + R$) and therefore we are still observing the epidemic in progress.

From this we can easily understand that the definitive defeat will only take place when we have proceeded to vaccinate a large part of the population, thus removing people from the S number and therefore *fuel* from the epidemic.

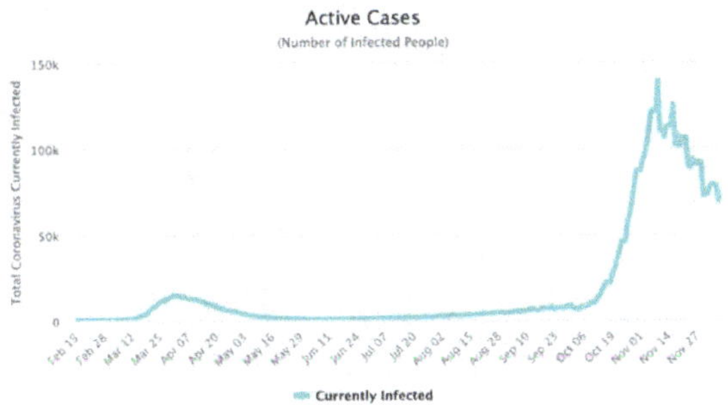

Figure 2: The graph shows the number of people currently infected in Switzerland. Data available on: www.worldometers.info.

In the meantime, what we can do is (unfortunately) to adopt lockdown measures that have the sole purpose of temporarily decreasing S (closing the population at home) and lowering the K factor, in order to take away speed from the spread of the epidemic.

In other words, the measures that the states are taking are aimed at limiting I, by acting on the speed of propagation (K) and the availability of S.

However, what happens once the lockdown is removed, is that the epidemic resumes because the virus is still in circulation and suddenly a large number of potential candidates for infection are again available. This is the origin of the subsequent waves: in the summer we had a decidedly low number of I (but not zero) because with the lockdown we had stolen petrol (keeping people under lock and then limiting the actual number of S) and using social distancing and masks (lowering K). Once the lockdown was removed, S went back to a big number and K rose at the same time. Hence the new peak of

September, which was tamed with a new lockdown implemented in October and which is still partly going on in these days.

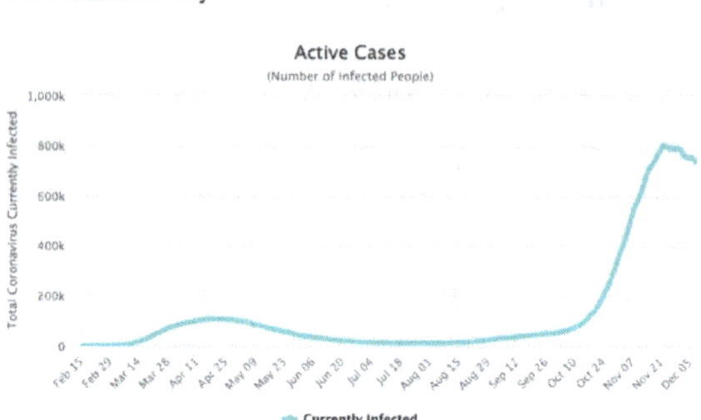

Figure 3: The graph shows the number of people currently infected in Italy. Data available on: www.worldometers.info.

What will happen in the future?

If this interpretation is correct, once the current lockdown is removed/mitigated, we will see sooner or later (January/February) the growth of a new peak of the infected and so on until the impact of the epidemic and, much more important, of the number of vaccinated, has not definitively subtracted people potentially capable of being infected, i.e. drastically decreased S. For this reason, it is very important, as soon as possible, to access a mass vaccination. If we don't, we will have successive waves again and again, with potentially very serious consequences.

Appeared in Italian in N. 51/2020 *L'Osservatore magazine.*

14. Unlikely swabs, effective vaccines and other accidents

Do not let the word *accidents* mislead you. We are not here to wish assorted misfortunes to anyone. Instead, in this episode of our column, we will talk about chance and probability, applied to the current pandemic. From the vocabulary, accident: *what happens by chance, random, fortuitous, unpredictable.*

But what do chance and probability have with the pandemic?

Answer: very much. For example, when we come across an infected person, then we have a certain probability of being infected. If we suspect that we are ill, we do a swab to find out if we have really been infected and have a certain probability of having a false positive result. If, on the other hand, we have been lucky enough to have already received the vaccine, we have coverage that works with the probability of 95%.

As you can see, in order to be able to speak with knowledge of the facts about infections, tests, vaccines and other aspects of this epidemic, we cannot help but bring up the theory of probability. A branch of mathematics that has taken its first steps in not exactly edifying environments. In fact, it is said that the first studies on probability and the role of chance, originated around gambling.

We are in the mid-seventeenth century and a French nobleman, the cavalier de Méré, turns to the most famous scientist of the time, Blaise Pascal, to solve a dice problem. From experience de Méré knows that the result of throwing a given is a random event and that it is easier to win if you bet on the release of at least a 6 in 4 rolls of a single die, than if you bet on a double 6 in 24 rolls of a pair of dice. According to his calculations, however, the probability (and therefore the frequency of the winnings) should be the same. How come experience contradicts his calculations?

Pascal explains the reason why and, by doing so, he starts the theory of probability. For those who want to know more, there is a clear explanation on the beautiful book by Boffetta and Vulpiani (Probabilità in Fisica. Un introduzione. Springer-Verlag Italy, 2012).

We stay in the company of prof. Vulpiani to see how probability also applies to the problems of the epidemic. Let's talk about tests: from swabs, to antigenic to serological; how to extricate yourself in the jungle of numbers? Do they work? Is it worth thinking about getting them?

In the book I wrote with Angelo Vulpiani on the *difficulty of predicting the future*, it tells of an interesting investigation carried out in the United States in the 90s of last century. Two leading psychology researchers did an experiment in which the following question was asked to a group of doctors and medical researchers at Harvard University: a disease has an incidence rate of 1/1000. There is a test that allows you to identify its presence. This test has a false positive rate (people who turn out to be sick on the test but are not) of 5%. An individual undergoes the test. The outcome is positive. What is the probability that the individual is actually ill?

More than four out of five of the participants answered that the probability was 95% (since the test fails only 5% of the time). Less than one in five of the participants replied that the probability was 2%.
Who was right?

Strange to say, the minority is right. With a test like this it doesn't make much sense to spend the money to get an answer that, if positive, doesn't guarantee that one is actually sick if not 2% of the time.
Here's how to think to get the right answer: suppose we apply the test to the Italian population, about 60 million. Since the disease affects 1/1000, it means that there are about 60,000 patients. If we test the entire population we will have a number of false positives of about 3 million. So, since you are really sick if you are one of the 60,000 then the probability of being sick is 60,000 / 3,000,000 =

2%. For the interested reader, I point out that the exact probability is calculated using the Bayes formula: PM / (PM + PE) where PM is the incidence rate of the disease and PE the probability of false positives.

This should be kept in mind as you rush to test for Covid-19. Let's try to make a calculation considering that currently the maximum incidence rate of the disease is around 1.3% of the population. Rapid antigen tests are notoriously popular because they allow for a response within about 30 minutes. Are they reliable? A recent study conducted by the Altamedica Research Center (https://www.agi.it/cronaca/news/2020-12-04/covid-test-antigenico-rapido-falsi-positiva-errori-10545007/) in Rome raises some doubts.

From the studies carried out it would seem that the probability of false negatives is in the order of 40%. This number would derive from the count of people who tested negative on the antigen test but then positive on the molecular swab (with less probability of false negatives). If this number is confirmed, then whoever takes the quick test and gets a negative answer means that he is only healthy with a probability of about 71%. Also in this case the count is simple: out of 60 million Italians, at the peak of the disease there were about 800,000 infected, so there were 59.2 million healthy so the probability of being healthy was PS = 59.2 / 60 = 98.7% and PE = 40% (false negatives). By applying the Bayes formula, the probability of being healthy after the negative test result is 98.7 / (98.7 + 40) = 71% approximately. Still high, but far from 100% certainty.

Based on what we have seen, according with the current spread of the disease, a test to be truly reliable must have a probability of false positives and negatives not larger than 5%.

There is also room for talking about probability when it comes to vaccines. There was a detailed debate on the efficacy values declared by the manufacturers: about 95% after 28 days from the first administration for the Pfizer and BioNTech product and slightly lower than that of Moderna (94.1%). AstraZeneca's pending approval is also rising from the initial 70% to 95%.

What do these numbers mean? Should we therefore be vaccinated? Answer: yes, as soon as possible. In fact, with a 95% effective vaccine we reduce our probability of getting sick drastically: if we currently have the 1.3% probability of being sick, once vaccinated our probability drops to 0.06%. Not bad for a little puncture on the arm.

Appeared in Italian in N. 3/2021 *L'Osservatore magazine*.

15. Much noise about nothing?

It has been at least since Shakespeare's time that the term noise takes on a negative character in our common understanding. In the famous comedy set in Messina, *Much Ado About Nothing*, the author uses the concept of *noise* to illustrate the useless and dispersive agitation of the protagonists, who pass from one misunderstanding to another, until the inevitable happy ending.

Even in common language, the term *noise* conveys an idea of disorder, something clearly unpleasant. In acoustics, *noise* is the equivalent of annoying sound, as opposed to music which, by definition, takes on the meaning of pleasant sound.

Things changed suddenly at the beginning of the last century, when Luigi Russolo, an Italian, amateur painter, sculptor and musician, who frequents Marinetti and the other Roman exponents of the Futurist movement, appears on the scene. In 1913 Russolo wrote a letter to his musician friend Francesco Balilla Pratella in which the role of noise in musical composition is exalted. The letter soon becomes a sort of futurist manifesto: *The Art of Noises*.

The French musicologist-semiologist Jean-Jacques Nattiez argues that it is formally possible to distinguish between musical sound and noise, by referring to the mathematical representation of a sound. To understand what it is, we must reflect on the fact that what we perceive as a sound is nothing more than the variation in the pressure that the air exerts on our ear and which induces vibrations on the eardrum (a membrane that connects the outer ear, to inner ear).

If such vibrations are of great amplitude, then the sound will appear to us as very intense. Conversely a weak sound will make the membrane vibrate only with small oscillations. Furthermore, if the vibrations are very frequent (the eardrum goes up and down many times per second) the sound will appear to us as acute tone, while it will be low tone if the oscillations are less frequent.

The role of vibrations as a source of sound can be observed on any membrane that is exposed to the air, such as the skin of a drum or the surface of a microphone. If we represent the amplitude of this vibration as a function of time, we obtain a signal of the type shown in figure 1.

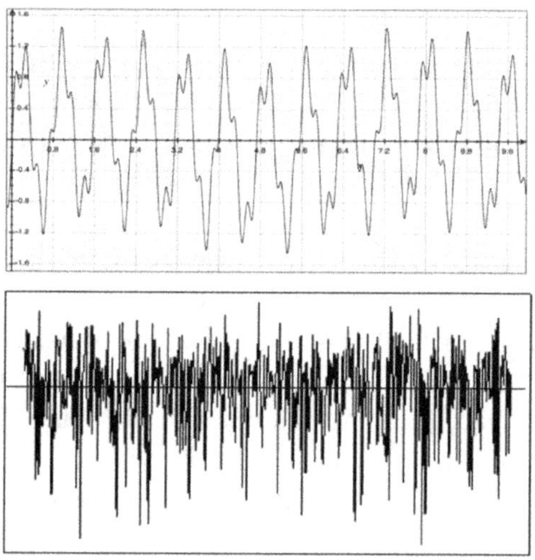

Figure 1. Top panel, a signal representing a sound (it is the sum of a finite number of periodic functions). Bottom panel, a noise with its characteristic messy appearance.

Here, two distinct vibrations are represented: the one at the top represents a musical sound while the one at the bottom represents a noise. How are they different?

To better appreciate the difference, we must resort to the work done at the beginning of the nineteenth century by a French scientist, Joseph Fourier. He discovered that all vibrations can be represented as a superposition of simple waveforms. Put simply, Fourier discovered that, however complicated it is, a vibration can be reconstructed by adding a number of elementary signals that are generally represented by simple mathematical functions. In the world of acoustics, it is common to use the trigonometric functions sine and cosine as simple mathematical functions.

Returning to Nattiez's distinction, we have that a musical signal can be expressed as the sum of a certain, limited number of regular periodic vibrations while for a noise this is not possible and we have to resort to an infinite sum of these functions. The result in the case of noise is that we no longer have an orderly and regular trend of the signal but a very messy one.

Once we understand the physical distinction between a musical sound and a noise, it becomes clear that musical sounds are in fact a very small and rather artificial class of all sounds, while noises occupy most of our acoustic experiences. Today we call noise, by analogy with sounds, also all those signals that are characterized by this high degree of irregularity. Remaining in the field of vibrations, the most important case in nature concerns what is known as *thermal noise*.

Thermal noise occurs in all material systems.
For example, the table on which I am writing this article and the computer or smartphone on which you are reading it, are vibrating right now. These are disordered vibrations of very small entity that we cannot perceive with our senses but nevertheless they are present and can be precisely measured with suitable instruments.

Once you realize this ubiquitous presence of vibrations, you understand how noise, far from being just an unpleasant sound, is actually a phenomenon that completely permeates our experience of the world. The reason why all things show this noise is related to the way in which thermal energy, or heat, manifests: e.g. in the form of random vibrations.
The surprising thing is that there is no way to stop these disordered vibrations unless you want to bring everything to the practically unattainable temperature of absolute zero.

One of the first scientists to notice the universal presence of noise was the Scottish botanist Robert Brown (in 1827), while he was observing, under a microscope, a grain of pollen lying on the surface of steady water. Due to the absence of any external perturbation, the pollen grain should have stood still, instead it vibrated continuously, disorderly moving in all directions.

Initially, Brown thought he had discovered a new form of microscopic life, but soon he had to change his mind: the motion continued even by replacing the pollen with light pieces of stone or wood. The random vibrations were a manifestation of the kinetic energy of the water molecules on whose surface all these objects were located: a particular form of thermal noise that we now call Brownian Motion.

To obtain a correct physical interpretation of this phenomenon, however, it will be necessary to wait until 1905, when Albert Einstein (yes, he again!) was the first to provide a theory in full accordance with experimental observations.

The history of noise is full of discoveries and interesting aspects. In my opinion, one of the most interesting concerns a discovery made at the University of Rome, La Sapienza, and whose fortieth anniversary falls in 2021. The name of this phenomenon in which noise, instead of creating disturbance, manages to improve things, is *Stochastic Resonance*. A truly fascinating story that we will tell in the next episode.
Stay tuned.

Appeared in Italian in N. 7/2021 *L'Osservatore magazine*.

16. The noise you don't expect

At first it is an almost imperceptible hum. Then, gradually, it increases in amplitude and expands in frequency, upwards. The effect obtained is remarkable. Some find it refreshing. Before, we used chill-out music for sound therapy but some customers pointed out that it was repetitive and a bit boring. So, we opted for noise. The popularity has increased considerably.

The director of the wellness centre of a luxury hotel tells me this with visible pride.

Noise is also this: a vehicle for therapy and potential well-being but also for possible stunning. A few years ago, the I-Doser mania was popular on the internet. A practice based on the use of audio files containing very low frequency noise recordings that would, if properly listened to, even have hallucinogenic power. How it is possible to listen to audio frequencies of the order of a few Hertz with the earphones remains a mystery, but this does not prevented some people from setting up a real business around the hypothetical drug via audio.

Although for a long time, as we saw in the previous essay, the concept of noise has often been associated with a negative connotation, in recent years a new understanding of noise and the role it can play in important physical and biological processes has appeared. Without thermal noise, for example, there would not be many of the phenomena occurring at the micro and nanoscale, typical of biochemical processes.

In the technological field, a positive conception of noise has gained some evidence and now there is the tendency of considering noise in conjunction with non-linear dynamic systems. It has been found that there are some interesting cases in which the simultaneous presence of noise and non-linearity can lead to situations of greater order and harmony, rather than disorder and disharmony. The typical example is the phenomenon of Stochastic Resonance, proposed for the first time in 1981 by Giorgio Parisi, Roberto Benzi, Alfonso Sutera and Angelo Vulpiani, at the La Sapienza University of Rome.

The story of this discovery is quite interesting and deserves to be told in detail. In 1981 the Roman physicists around Parisi were interested with many different and unrelated topics, one of these was the attempt to explain why the ice ages or, more precisely, periods of below-the-average earth's temperature, are repeated with a surprising periodicity: about every hundred thousand years. During the cold phase, the temperature drops by about 10 degrees on average. This decline would seem to be due to a lower insolation of the earth due to a different angle of exposure of its surface, following a periodic oscillation of the earth's axis. The puzzle physicists faced was that, with their two-state model (cold earth and warm earth), the perturbation due to the oscillations of the earth's axis was too small to cause this transition from one state to another.

Giorgio Parisi, at that time the group leader, intervened in the discussion suggesting that perhaps they should have taken into consideration the effect of temperature fluctuations, due to other accidental causes. These would have possibly provided the energy necessary to allow the change of state. The suggestion, seen by a dynamical systems specialist, appears rather unusual. In simple terms, it is like if someone suggested that if you can't hear the conversation with your desk neighbour, to improve things it would be a good idea to open the window and let in the noise of the nearby traffic.

It is hard to think that the presence of external noise can improve your understanding of the words of your interlocutor. And in fact, that's usually the case. However, there is a particular class of physical systems (bistable systems) for which, instead, in a very surprising way, if you add noise, the physical system, instead of increasing the degree of disorder and confusion, begins to behave in a more orderly way.

Said out of metaphor, in the case of glacial periods, the addition of noise had the purpose of encouraging periodic transitions from one state to another. This behaviour was immediately very surprising and absolutely counterintuitive.

The article published by the Roman colleagues was widely disseminated and the phenomenon, which they called "Stochastic Resonance", was soon assumed as a paradigm of all cases in which the addition of noise, rather than deteriorating the performance of a system, can on the contrary, improve functioning.

The logo of the International Conference for the fortieth anniversary of the discovery of Stochastic Resonance. To be held in Perugia from 13 to 15 Sept. 2021.

Another case, similar to the *Stochastic Resonance*, is the phenomenon of *dithering*.

It is often narrated that the first mechanical computers installed aboard British aircrafts, during the Second World War, and used to calculate the coordinates of the release of bombs, worked much better in flight than when they were on the ground. The cause was soon attributed to the random vibrations that during the flight came to "disturb" the functioning of the computers. Instead of disturbing them, however, the vibrations instead caused a more regular functioning. Also in this case, as for Stochastic Resonance one, the reason for the positive effect of noise lies in its interaction with a non-linear physical system. In the case of the glaciations it was the bistable system (hot-cold), in the case of the computers it was the switches (closed-open) that made the calculation operations proceed.

To celebrate these surprising and beneficial aspects of noise, on 13-15 September, on the occasion of the fortieth anniversary of 1981, we will hold an international conference in Perugia dedicated to Stochastic Resonance and all these surprising phenomena.

Stay tuned, because we're not done with the noise yet: the next episode will show you how to turn noise into energy and get rid of the hateful electric batteries once and for all.

Appeared in Italian in N.11/2021 *L'Osservatore magazine*.

17. Beyond the Standard Model, in search of a New Physics

Among the recent scientific news, an announcement that comes from the Fermilab laboratory in Chicago has made some noise.

To understand what it is, we have to go back a bit in time: about a hundred years ago. Many of us, with a basic scientific background, surely know what Richard Feynmann considered the most important scientific statement worth passing down to future generations:

all matter is made of atoms.

Each atom is composed of three types of *elementary* particles: Protons, Neutrons and Electrons. The exception is the Hydrogen atom, the smallest of all atoms, which has only one proton and one electron.

About a hundred years ago, when recently even the last sceptics had finally accepted the existence of atoms, it was realized that, in reality, there were more than three elementary particles. In fact we discovered that they weren't elementary at all, given that during the collisions between them, they could break down into other new particles.

These new particles have exotic names, mostly inspired by the Greek alphabet. Among these is the Muon, discovered in 1936 by Carl D. Anderson and Seth Neddermeyer at Caltech, while they were studying the so-called cosmic rays, other particles that literally rain down on the earth from space.

As the discoveries proceeded and new experimental devices were built, such as the accelerators at CERN in Geneva, the list of particles enriched with new particles, all different in size, electric charge, mass and other physical properties. It took nearly fifty years to put this jumble of discoveries in order. What emerged is a complicated but very elegant picture that physicists call the *Standard Model* (Figure 1).

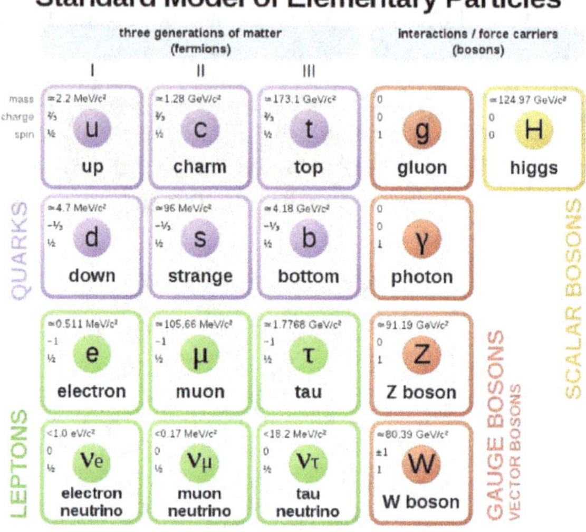

Figure 1. Summary table of the elementary particles of the Standard Model (source Wikipedia).

The standard model represents the synthesis of what we know today about the microscopic structure of nature and is one of the most elegant and complete theories we currently have in physics.

It takes into account Einstein's special relativity and all quantum mechanics. According to this theory, everything we know in nature originates from the interactions between two types of elementary components: the particles of matter, called *Fermions*, in honour of the Italian physicist Enrico Fermi and the particles of force, called *Bosons*, in honour of the Indian physicist Satyendranath Bose.

In turn, the particles of matter are divided into *Quarks* and *Leptones*. Quarks are the elementary components of the most familiar Neutrons and Protons that we all know, while among the Leptones there is the very popular Electron we deal with every time we turn on the light switch.

Speaking about light: as is known, this is a manifestation of the electromagnetic force which is one of the four fundamental forces of the universe (the other three are *gravity*, the *strong nuclear force* and the *weak nuclear force*). Among the particles of forces there is therefore the *Photon* which is precisely the particle of light. Along with him there are also other particles including the recently observed particle called the *Higgs boson*.

It is important to keep in mind that the Higgs boson was predicted by the English physicist Peter Higgs in 1964 and observed only in 2012 in the ATLAS and CMS experiments, conducted with the LHC accelerator at CERN. At the moment of its actual observation (April 5, 2012, in the 27 km accelerator, under the border between Switzerland and France) the physicists exulted with satisfaction but our knowledge of the plot of the world, in reality, has not advanced much. And this is not because it was not a great discovery, but because the evidence of the existence of this boson did not add much new to what we knew, this having been a long-awaited result and perfectly in line with the standard model.

Physicists are somewhat strange scientists: first they perform many experiments to try to find a theory that explains the results obtained in the experiments. When they succeed, after eventually having taken some Nobel Prize, they continue to make new experiments in the hope that these will give results that are not explainable by the theory so laboriously obtained. If they do not obtain contradicting results, there is no fun, because it means that the existing theory is still valid and there is nothing new. This is somewhat of what is happening with the Standard Model; the experiments of the last twenty years have done nothing but confirm its validity and those in search of a "New Physics" have so far been disappointed.

The news that has appeared in recent days goes precisely in this direction: it would seem (and I say it *would seem*) that an experiment conducted at Fermilab, aimed at measuring the value of the magnetic moment of the Muon, gives results that may differ from those predicted by the Standard Model. The deviation is still not very significant from a statistical point of view but it is a possible clue that there is something to fix in the theory.

To be more precise, the experiments are aimed at measuring the so-called "g-2" factor. In the previous theory, due to the great English physicist Paul Adrien Maurice Dirac, the g factor that appears in the value of the magnetic moment is exactly 2. Thanks to the Standard Model we realized that there are "anomalies" due to contributions to this value that derive from the existence of virtual particles. These are particle-antiparticle pairs that populate the vacuum and, as it were, make it simmer due to their random creation and annihilation processes. An antiparticle is just a particle with the opposite charge. There is the electron and the antielectron, which has a positive charge. Here, this boiling of the vacuum is one of the causes why the factor g is not exactly equal to 2. The Standard Model provides a certain value of "g-2" which, however, in recent American experiments would differ from that observed for less of one part per million. The deviation is very little really, but it could be enough to make us understand that the theory is incomplete and we need new ideas to be able to do better.

Appeared in Italian in N. 17/2021 *L'Osservatore magazine.*

18. Vibrations and that Swiss invention of long ago

I keep my promise and go back, for one last time, to talk to you about noise. This time we are dealing with energy, that mysterious quantity that goes into running cars, warming us up in cold weather, running our washing machines and, ultimately, keeping us alive.

The issue of energy has become of general importance especially since the oil crises of the 1970s, when the change of power in some of the oil-producing states caused a sharp decrease in the availability of crude oil and therefore a rapid increase in oil prices. For the first time in 1973 and then, even more particularly in 1979 with the Iranian revolution, the public opinion of Western countries was forced to deal with physical concepts like energy, efficiency, friction, dissipation and other amenities that until then had remained confined inside laboratories of physicists and engineers.

The concept of energy was first introduced during the so-called first industrial revolution, in the context of thermodynamic studies, by people such as Leibnitz and Young. In the early nineteenth century, the French scientist Gustave-Gaspard Coriolis in 1829 introduced the name *kinetic energy* for the first time, to indicate an energy associated with movement. Everything that moves has an energy that is greater the greater its speed. This energy can actually be used to do work and, once transformed into electricity, it can be easily used instantly or stored in batteries for later use.

The energy we use to recharge our mobile phones sometimes comes from movement, for example the water that falls in a waterfall and sets the turbines in rotation or the rotation of the arms driven by the wind in a wind turbine.

Figure 1. 1778 automatic watch with rotating weight. Mazzi firm, Locarno. (source Wikipedia).

So, if movement is a form of energy, why not use the vibrations caused by noise, which are certainly a form of movement, to produce energy? The idea, daring as it may seem, is not new. Remember the wristwatch that recharges with the movement of the arm? Well, the first to show a working prototype of the self-winding watch was a Swiss inventor (not for nothing!), Abraham-Louis Perrelet, who lived in Le Locle in 1776. The invention was recognized the following year from the Genevan society of arts, which declared that a 15-minute walk was enough to recharge the wristwatch. Since then, a long way (literally) has been travelled and the idea has been applied to devices other than the clock.

About thirty years ago, a new field of research was born, called *Energy Harvesting* (EH). The idea behind the EH is the following: for portable electronic devices (such as a modern wristwatch) you must not resort to batteries that have a whole series of troubles, starting with the fact that they run out of energy, are bulky, heavy, expensive and pose problems for their disposal. No, we need to embrace a different approach:

use the energy available in the environment and transform it into electricity for immediate use.

In this sense, the transformation of vibrations into electrical energy constitutes, so to speak, the mother of all solutions. Since the world is full of vibrations, all that remains is to arm yourself with good will and build systems that transform this kinetic energy into electrical energy, for example using piezoelectric materials, and use it to power mobile electronic devices. So far so good, except that when one passes from words to deeds, problems arise.

The first and most important problem is that vibrations are usually very different from the movement of an arm on which a watch is placed on the wrist. The oscillations of the arm are quite periodic while the vibrations, by their nature, are random. We have again here the periodic versus random distinction that we encountered when we talked about sound and noise a few episodes ago.

To wind a watch only with vibrations, much more energy is needed because this, instead of being concentrated on a single frequency is instead distributed over a wide spectrum of frequencies, as usually happens for noise. For this reason, it has been much more difficult to find efficient solutions for energy production.

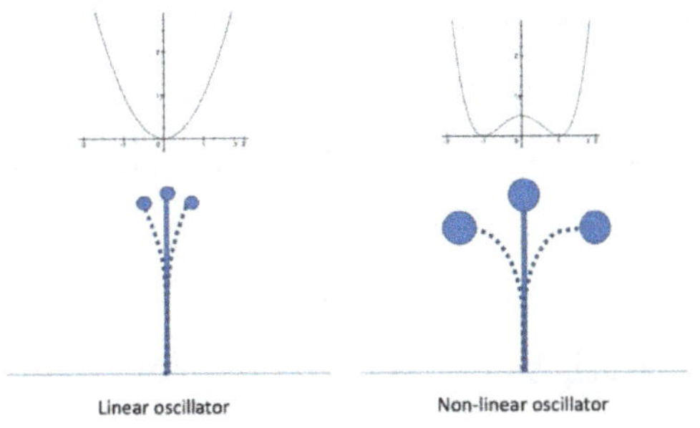

Figure 2. Linear (left) and non-linear (right) oscillations.

However, the situation has changed in recent times, when it was discovered that, instead of using a linear oscillator, such as the mechanical rotor in Figure 1, it was possible to choose a much more complicated but also much more effective type of oscillator, which we will call non-linear oscillator. This discovery was made about fifteen years ago in our NiPS (Noise in Physical Systems) laboratory at the University of Perugia.

To give you an idea of what a non-linear oscillator is, you can imagine a flexible rod planted on the ground at the upper end of which a certain mass is applied (Figure 2). If the mass is small, we will have linear oscillations (which can be represented with a parabolic motion, as in the figure on the left above, where we have only one point of equilibrium, placed in the centre). If, on the other hand, the mass is quite large, the oscillations will tend to be non-linear, with two equilibrium points on the right and left (Figure 2 on the right).

From the discovery that a non-linear oscillator, made of piezoelectric material, could transform vibrations into electrical energy much more effectively than a simple linear oscillator, a real company was born (a spin-off of our laboratory) called Wisepower srl (www.wisepower.it). This company today produces devices for the structural monitoring of viaducts, telecommunication towers, oil pipelines,... which work without batteries and use the vibrations present in these structures to obtain the electrical energy necessary for their operation.

Not bad for a simple noise that was supposed to be just an annoying sound.

Appeared in Italian in N. 21/2021 *L'Osservatore magazine.*

19. Atoms and bits

We live in the era of *digital culture*. We are told so every day by any kind of media (mostly digital themselves). The Mr. Draghi government in Italy has created the recent ministry of *technological innovation and digital transition*, one of the strengths of its program.

In Davos, on January 21 2021, Accenture presented to the elite of the world executive class, a study entitled *The European Double Up: A twin strategy that will strengthen competitiveness* which states that companies that will be able to accelerate the transition to digital and sustainability will be able to emerge stronger from the Covid-19 crisis(https://www.accenture.com/us-en/insights/strategy).

Digital culture is the product of a technology born in an age when noise was an enemy and electronics were still in the cradle. In the early 1940s, sending a message was not an easy task and the quality of the one received often differed considerably from the original, due to the unwanted action of noise. The natural solution was to reduce the variability of the signal, caused by fluctuations, by imposing a finite number of amplitude levels. Clearly, the most efficient reduction was to limit to just two different levels: what we now call binary representation.

It is important to remember that this reduction process, technically called digitization, i.e. the action aimed at transforming a continuous (analog) signal into a discrete (digital) one, is an operation that is not free from errors, so that the digitized signal always differs from the original one. Despite this limitation, scientists have used digitization to build tools that have led to undeniable social progress.

Undoubtedly, the most popular of these tools is the digital computer. In addition, we have had digital communication systems, such as digital telephones, digital radios and digital TVs, and also digital industrial machines. To carry out their tasks, these digital tools need input data that is obtained from the physical world through digitization procedures, performed by special devices called ADCs (Analog-to-Digital Converter). The large amount of data available has raised some concern about their ownership and management (see, for example, "Dated from birth. A threat to freedom", Claudia La Via, Wednesday 24 February 2021, Avvenire).

The impact of digital technology has affected not only the way we work and communicate, but also our understanding of the world around us.

Ralf Landauer (1927-1999) first advanced the idea that the amount of information stored in a computer's memory could play a role in the physics of the device itself, being interpretable as part of the computer's entropy. Although this position has long been debated in the physics community and I myself, among others, have worked to prove that its conclusion was wrong (Computing study refutes famous claim that 'information is physical'- https://phys.org /news/2016-07-refutes-famous-physical.html), the application of the information theory paradigm to the interpretation of the physical world, has become widespread.

Overall, there is no doubt that digital technology has revolutionized our society by favouring a progressive lightening, through a widespread process of dematerialization. Digital music, for example, is immaterial and is discharged as electromagnetic radiation into portable devices weighing a few grams. Dedicated digital tools have also dematerialized newspapers, films, books, paintings, sculptures and, of course, money.

This progressive weight loss did not go unnoticed. One of Italo Calvino's famous "American Lectures", probably the best, is dedicated to *lightness*. This lecture was to be presented for the Charles Eliot Norton Lectures at Harvard in the fall of 1985, but was never given because Calvin died suddenly on September 19 of that year. Calvino writes:

my working method has often involved the subtraction of weight. I tried to remove weight, sometimes from people, sometimes from celestial bodies, sometimes from cities; above all I tried to remove weight from the structure of the stories and from the language .

The work program announced by Calvino was taken very seriously by those who, in the following years, designed and distributed the digital tools that we use so much today, with the unexpected result of overwhelming us with an enormous amount of digital data.

The dematerialization process we have witnessed over the past forty years, was aimed at removing information from atoms and transferring it into bits. In my humble opinion it's time to change things. If we want to stop the deluge of data and the risks associated with it, if we want to preserve our democracies and protect our freedom, it is time to reverse the process and defend the reasons of the atoms against those of the bits. Surprisingly, some first steps in this direction have already been taken by the pioneers of 3D printing. In a 2008 book titled *Fab: The Coming Revolution on Your Desktop - from Personal Computers to Personal Fabrication*, Neil Gershenfeld, director of the MIT Center for Bits and Atoms, discussed the potential of new generations of nerds interested in making objects: real things, not virtual. A clear sign that the revenge of the atoms has begun.

What will the future of digital culture be then?

To answer this question, Italo Calvino also comes to our aid. In the same lecture on lightness, Calvin speaks of the important legacy of the Roman poet Lucretius (c.99-c.55 BC) and quotes a beautiful verse from his description of the material world:

the specks of dust swirling in a ray of sunshine in a dark room

(II.114-124).

What a beautiful description of something that is light without being empty. What an unlimited source of information there is in this vortex of dust, impossible to contain in a finite digital representation and yet so fundamental in our description of reality.

Fluctuation is the word that comes to mind when I listen to such a description. Fluctuation and noise - exactly what we wanted to get rid of at the beginning of the digitization process. While it is true that the future belongs to those who can imagine it, we need more noise and fluctuation to grasp the unsimplified reality we live in. No digital representation can ever embrace it. Analog technology will eventually take the place of digital technology and we will need a new generation of scientists.

Appeared in Italian in N. 25/2021 *L'Osservatore magazine.*

20. Memory doesn't last forever

Anyone who has had the direct experience of a person suffering from Alzheimer's or short-term memory loss syndrome (anterograde amnesia), knows what difficulty one must face even in conducting the most trivial daily activities. Memory represents for humans a function of fundamental importance. However, no matter how important the object of memory is, even in the absence of particular pathologies, it is a common experience that memories tend to weaken over time.

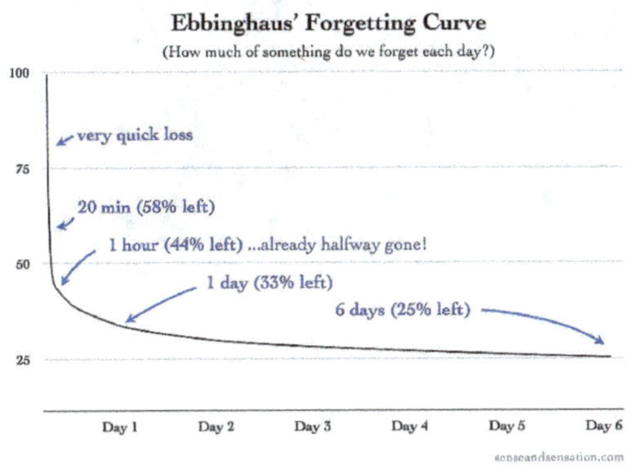

Figure 1: Hermann Ebbinghaus. Forgetting curve. (Source: Wikimedia Commons)

Probably the first who systematically studied this phenomenon was the Prussian philosopher Hermann Ebbinghaus (1850-1909), who conducted a series of studies on the persistence of human memory, collected in an 1885 text entitled *Memory: a contribution to experimental psychology*. Ebbinghaus is credited with discovering the so-called forgetting curve, an exponentially decreasing curve that shows that the ability to remember, unless continuously reinforced, tends to halve at regular intervals.

Given the natural weakness of our ability to remember, the passage from the oral tradition to the written tradition constituted a stage of development of enormous importance in the formation of human culture. For the first time, the wealth of knowledge that humanity had accumulated in previous generations could be spread over time, with safety and without limits of breadth. It is no coincidence that the people of Israel are identified with the people of the Book.

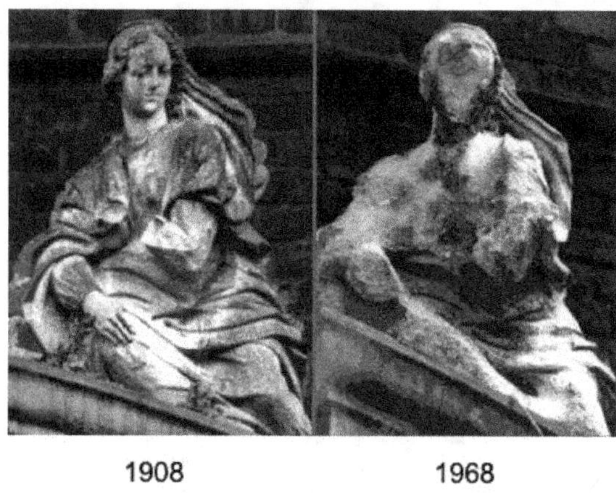

Figure 2: The same monument photographed sixty years later.

However, even books can perish and we still mourn the inestimable loss of most of the papyri in the library of Alexandria. If we think about it, we realize that even the material objects, artefacts, monuments and landscapes to which we have entrusted our sign, require periodic maintenance to counterbalance the effect of time which modifies the differences in shapes and levels (as in figure 2). If there had not been a continuous restoration activity, the great masterpieces of the past would have been cancelled and the memory they embody, irretrievably lost. This tendency to erase memory is nothing more than a manifestation of the second principle of thermodynamics which, in a system left to evolve spontaneously, produces an inevitable increase in entropy and therefore a loss of details and differences.

The *digital transition civilization* we are building is not immune to this phenomenon. In digital devices the information is usually stored in the form of binary numbers, where the bit "0" has the same value as the bit "1", and it is customary to use an abstract representation of such devices in terms of dynamic systems with two stable states equally energetic, separated by a potential barrier (see figure 3).

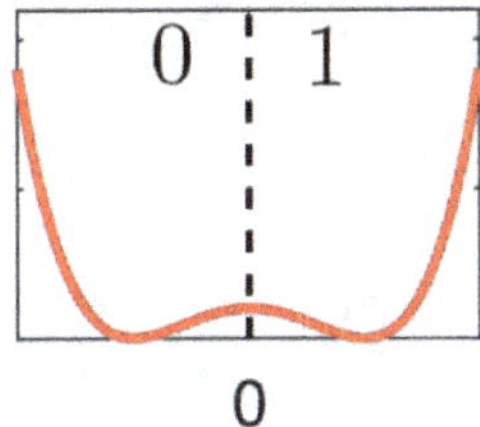

Figure 3: Dynamic representation of a digital memory.

The potential barrier separates the two logic states ($x < 0$, which represents the bit "0" and $x > 0$, which represents the bit "1") and confines the dynamics within one of the states, as required by the memorization. This is clearly a non-equilibrium condition that evolves, within the relaxation time of the system, towards thermal equilibrium (represented by the condition of same probability of being in one state or another). During this process, storage errors accumulate to the point where the stored bit is statistically lost.

To combat the loss of information, it is therefore customary to set up a *refresh procedure* that acts against the relaxation process at equilibrium. In a recent work (D. Chiuchiù, M. López-Suárez, I. Neri, M. C. Diamantini and L. Gammaitoni; Fis. Rev. A 97, 052108, 2018), at NiPS (Noise in Physical Systems) laboratory, we studied the refresh mechanism and its consequences for the preservation of digital memory.

The conclusions we have reached are interesting and, in some ways, unexpected.

The **first** conclusion represents good news: it is possible to preserve a memory for a given time with a given probability of error by spending even a very small amount of energy, as long as the refresh operation is performed very often or very close to the equilibrium condition.

The **second** conclusion is not as comforting: no matter how much energy we invest or how often we refresh, there is a maximum time after which all memory will be irretrievably lost.

This second conclusion is the consequence of a very fundamental physical principle: the Heisenberg uncertainty principle, which imposes the existence of a minimum non-zero uncertainty, during the writing operation. This uncertainty ends up growing over time and eventually affects the existence of the memory itself that is thus irretrievably lost. After how long? It depends on the type of memory and the error we are willing to accept. For example, for certain types of memories, at room temperature, an acceptable probability of error of one part per million, will give a maximum life of 2 years. If we are willing to accept a probability of error equal to one part per ten thousand, the maximum duration will be 200 years.

The existence of a maximum duration time for each possible storage medium could be surprising. We are in fact used to ancient artefacts and paintings that constitute the memory of past times. The law of the finite duration of memory, of every memory, also applies to them. They are analog memories that record information encoded by other men in other times.

Each form of memory has its necessary refresh period and its maximum duration time, depending on the specific nature of the material medium in which it is made. It can range from a few seconds, for electronic digital memories if no updates are done, to (probably) millions of years for carefully preserved monuments.

Will we be able to preserve the large amount of digital data our society is producing? And if so for how long?

If we decide to build a future society on digital data, then it's best to be aware that it won't last forever - the more volatile and error-prone our memories are, the shorter their lifespan.
The same will happen to our society.

Appeared in Italian in N. 29/2021 *L'Osservatore magazine.*

21. If data is the new oil, who are the new oilmen?

"Data is the new oil" is a phrase that we hear often in these times.

It was coined in 2006 by Clive Robert Humby, an English mathematician and entrepreneur, expert in what we now call *data science* and which only a few years ago we would have called *statistics*. In fact, for many decades, statistics have been the science of data, or the mathematical discipline dedicated to data analysis, whatever their nature. From personal data, for the analysis of which IBM was founded in 1911, to economic data, which is why statistics are considered the mathematics of economists.

For some years now, however, a new trend has emerged that tends to associate data analysis with Artificial Intelligence (AI). AI (which we talked further back in this book) has made significant progress only in recent years by introducing computational techniques based on a relatively new concept: the neural network.

The idea behind this new concept is relatively simple to explain. An artificial neural network (inspired by how the brain is structured) is a set of nodes connected in a more or less random way. Each node can be in an "on" or "off" state in relation to what are doing the nodes it is connected to. In a network there are input nodes and output nodes and, depending on how the switching "on" or "off" rules are established and how much the individual connections weigh, calculation algorithms can be coded that provide the result in the output nodes in function of the value of the input nodes.

Using this type of schemes, it is possible to implement the computing technologies known as *deep learning* at the basis of current AI. We have talked about them previously and we have seen how these are essentially very efficient classification schemes, successfully used for image recognition and, subsequently, for speech recognition.

These powerful techniques, in order to work, need data, a lot of data. For example, to "train" an image recognition algorithm containing a cat, we need a great many (tens or hundreds of thousands) of images that we already know contain a cat. These images are used to create the artificial neural network that will then have to implement the *deep learning* algorithm. Similarly, in other sectors it is necessary to have a lot of data to create the algorithms that will be used (hence the expression *Big Data*).

Probably the first case of international relevance in the application of these algorithms is linked to the scandal of the English company *Cambridge Analytica*. In 2018 the company was the focus of a judicial investigation for using personal data, acquired by Facebook, to build a voter profiling at political events such as brexit and US presidential elections.

Since then, many have realized that data has the potential to be of great value even though, as Michael Palmer of the Association of National Advertisers noted, *Data is just like crude. It's valuable, but if unrefined it cannot really be used.*

So, we come to the initial question: if data is the new oil, who are the oilmen?
Just as the old oilmen are those who extract crude oil from the ground and then refine it, the new oilmen are those who are capable of capturing the data and analysing it to obtain useful predictions.

After the experience of Cambridge Analytica, the owners of the oil fields (Facebook, Amazon, Google, ...) followed the example of some Arab countries: they hunted foreign oilmen and began to extract and refine their data by themselves. One of the most interesting applications of this new policy is the so-called "predictive marketing".

We experience this technology whenever we receive advice via email or on the web about the purchase of a boat (after having visited sites and conversations related to boats in the previous days) or a special cream against sun burns (after having visited sites on a dermatological and beach theme).

In all this we must not forget a fundamental aspect: who produces the drills? In fact, oil cannot be extracted without drilling and those who produce the extraction technologies have a considerable advantage over all aspiring oilmen. Apple, Samsung and the like have so far been the producers of modern drilling rigs for data acquisition and in fact they too have begun to grow oil ambitions (think of the subtle invasiveness of *Siri* or *Alexa*).

For the future? In the future, scenarios of great interest are opening up for all those who will be able to produce new types of drills that go beyond the smartphone. It is the vast prairie called **IoT** (Internet of Things) where it will no longer be just humans who will produce data, but mostly things. Think of the vibrational sensors inserted in the infrastructures of a motorway viaduct capable of measuring the structural health of the bridge and measuring the traffic load, just to give an example that is already a reality. The black gold rush has already begun, those who want to take part are warned: we are in the new far west, saddle your horses.

P.S. As I had just finished writing this article, I received an email from a German consulting company. They ask me, for one of their clients, if my laboratory is able to devise new sensors to be inserted in cars in order to monitor the health of passengers without coming into contact with them. They are interested in sensors that detect temperature, blood pressure, heart rate, breathing rate, oxygen saturation data.
As we said ... the race has already begun.

Appeared in Italian in N. 34/2021 *L'Osservatore magazine*.

22. Surveillance Capitalism and the future technology

There is no doubt that the introduction of automatic calculation has been of enormous importance in our society. Today's culture has been shaped, at least in the past 70 years, by computers and their technology.

What will our near future look like?
Answering this question is both necessary and difficult.

It is necessary because, as an old adage attributed to Abraham Lincoln goes, *the best way to predict the future is to create it*. We would like to live and let our children live in a world that, at least a little, we have helped to shape.

It is difficult because, if we try to look back only 70 years, up to the 1950s, we realize that it must have been really hard to imagine in those years what the impact of computer-related technology would have been. For example, in the literature of the time there is no trace of the Internet or of smartphones, with the exception of Dick Tracy, an American cartoon drawn by Chester Gould which appeared in the early 1930s as a strip in the Detroit Mirror, starting from the end of the in the 1940s. He equips himself with a wrist videophone.

The enormous transformative action of the so-called digital culture (we talked about it in the piece entitled *Atoms and bits*) has surprised many contemporary intellectuals, also arousing some concern. Among the most concerned observers is certainly Shoshana Zuboff, a professor at Harvard Business School, who on January 19 last published an article in the New York Times significantly entitled *The Coup We Are Not Talking About*. The piece is inspired by a famous book by her entitled *The Age for Surveillance Capitalism. The Fight for a Human Future at the New Frontier of Power* (Profile Books, 2019).

In this article, Zuboff raises a serious alarm about the future of Western democracies:

"Our time is comparable to the early era of industrialization, when owners had all the power, their property rights privileged above all other considerations. The intolerable truth of our current condition is that America and most other liberal democracies have, so far, ceded the ownership and operation of all things digital to the political economics of private surveillance capital, which now vies with democracy over the fundamental rights and principles that will define our societal order in this century."

The point raised by the author is that the dizzying development of technology in recent decades has, in fact, shifted control of the levers of society into the hands of a few large industrialists, whose names are quite familiar to us: Bill Gates (Microsoft), Tim Cook (Apple), Larry Page (Google), Jeff Bezos (Amazon), Mark Zuckerberg (Facebook).

Is this alarm justified?
If we think of the great pervasiveness of digital media, we suspect that the control of the contents conveyed through them, may actually play an important role in the orientation of public opinion. The scandal created by the British company *Cambridge Analytica*, accused of having manipulated information disseminated by the media for

electoral purposes during the US campaigns and the Brexit referendum, has brought the problem of digital media control to everyone's attention.

As if that were not enough, the uncontrolled spread of fake news that has made the Covid-19 pandemic even more lethal, reminds us that the danger has not been tamed.

At the base of the great impact of digital media is a technological development that is increasingly in the hands of a small circle of industrialists interested in making, as is natural, the interests of their shareholders first and then, perhaps, those of the rest.

In my opinion, the way out of this potentially dangerous condition is not, as many hope, in the introduction of strict constraints on the way in which social media are used, but rather in a double action: the recovery of public initiative in the field of research and the promotion of widespread digital entrepreneurship. I believe it is important, after years and years of uninterrupted cuts, to significantly fund public research carried out at universities and national and international research centers. At the same time, it is necessary to update the educational offer of universities to intercept the new demand for use of the enormous amount of digital data available.

Certainly "data science" courses should be established in technical-scientific faculties but, at the same time, appropriate courses for degrees in Political Science should be promoted, so that the consequences of this recent technology on the functioning of our society can be explored.

The other category of actions concerns the promotion of widespread digital entrepreneurship, financing the birth of innovative start-ups with public money, perhaps inspired by research developed at universities. For years there has been talk of start-up promotion policy but it is hard to see the fruits of the efforts made so far. One of the reasons lies in the perverse logic of the co-financing mechanism employed. Just to create a start-up made up of young graduates from scratch, at least 100,000 euros are needed. The incentives from the government reach up to 50,000 euros (after replying to a specific call) and this implies that young entrepreneurs have to invest the other 50,000 euros. Well, according to the writer's experience, young researchers almost never have the money available to match public investment, with the result that even brilliant ideas remain in the drawer.

In conclusion, to avoid surveillance capitalism it is necessary to change policy towards research and its economic repercussions.
And this is a good time to do it.

Appeared in Italian in N. 39/2021 *L'Osservatore magazine.*

23. Physics: A quality Nobel

It is very likely that, in the not too distant future, looking at our age, it will be remembered as the age in which the concept of *quantity* reigned supreme over all other criteria, including *quality*. And I am not referring only to the universally widespread habit of accumulating money or measuring the value of every person or thing in purely economic terms, nor to the quantitative indices we use to determine the wealth of a country or continent. If you think about it, you will find that the *quantitative* aspect permeates many of our daily judgments.

Isn't it true that teenagers are fascinated by influencers who can boast millions of followers? This, of course, without any regard for who these subjects actually are: humans, chatbots...
And isn't it true that among the social media goers like Facebook, the success of posts is measured in the number of likes or shares? Again, with no control over who or what is "liking" or sharing the post.

Quantity is a very popular and almost undisputed yardstick. It is also for this widespread situation, which otherwise would be an occasion of discouragement for me, that I am pleased to see that the Nobel Prize in Physics, this year, was given to a scientist who paid attention not only to the quantitative aspects, but also, to the qualitative aspects of the reality of the physical world. Giorgio Parisi, in fact, was awarded for his studies on complexity, with particular reference to the discovery of the interconnection between disorder and fluctuations in physical systems.

Complex systems are those physical systems, to put it briefly, in which the overall effect is very different from the sum of the effects of the individual parts of which they are composed. It is equivalent to saying that they are systems for which a mere *quantitative* criterion of addition is not sufficient to explain their behaviors.

Let's take an example.

If we study the motion of a single H_2O water molecule within a liter of liquid, we could obtain useful information which, however, will never be able to explain the splendid phenomenology of the vortices that form in a stream. A vortex is a complex structure composed of a very large number of water molecules, in which the overall behavior cannot be simply explained as the sum of the behaviors of single molecules.

From left to right: Luca Gammaitoni, Giorgio Parisi.
Photo taken by Paolo Tosti at the conference on Stochastic Resonance, held in Perugia in August 2008.

Another example is the climate, a complex system par excellence, in which the overall behaviour cannot simply be traced back to the dynamics of the individual gaseous molecules that make up the atmosphere. To describe the motion of a fluid in a stream or of a gas in the atmosphere, it is necessary to introduce a description that takes into consideration structures that appear only if we look at the behaviour on a larger scale than that of its components. In short, this description requires a rigorous mathematical (and thus *quantitative*) treatment of the laws of Physics, but it cannot ignore a look that is also *qualitative* on nature.

We have already spoken several times in this column about the important role of fluctuations in the description of the behaviour of complex systems. Last time it was in the issue of March 13, 2021, in which we described the phenomenon of *Stochastic Resonance*, discovered by Giorgio Parisi, together with Roberto Benzi, Angelo Sutera and Angelo Vulpiani, at the *La Sapienza* University, in Rome, in the early eighties of last century. With a little luck we were able to anticipate the Academy of Sciences in Stockholm, because it is also for the discovery of the *Stochastic Resonance* that Parisi was awarded.

In this case it is the counterintuitive role of the fluctuations which, instead of being a source of confusion, contribute to increasing the condition of order (a *quality*) of the system, which must be carefully considered.

It is no coincidence that the other two winners of the Nobel Prize, the Japanese naturalized American Syukuro Manabe, and the German Klaus Hasselmann, also deal with climate models, a sector where fluctuations are the master. How can we forget Ian Malcom, the scientist played by Jeff Goldblum in the first Jurassic Park, who presents the theory of CHAOS by telling the story of the butterfly that flapping its wings in Brazil could change the climatic conditions to the point of causing a tornado in Texas? Small fluctuations which, thanks to the complexity of the climate system, can be amplified to even become a tornado.

The physics of complex systems is surprising, not easily predictable and requires qualitative approaches rather than merely quantitative approaches. And there is still a lot to discover: any volunteers?

Appeared in Italian in N. 44/2021 *L'Osservatore magazine.*

24. The Omicron variant and the evolution of the pandemic

After many months we return to talk, in scientific terms, of an epidemic in this column. Driven on the one hand by the desire to finally see this terrible pandemic end and, on the other, by the need to know what the effects of the new Omicron variant will be, we ask ourselves the crucial question: how the Covid-19 pandemic, which is still underway, will evolve. ?

Answering this question means exercising that difficult art of predicting the future from which, in some way, this column takes its name (*traces of the future* in L'Osservatore Magazine).

To follow the traces of the future that the present pandemic condition is disseminating, we must start from the mathematical models we use to study this type of phenomena. In a previous issue we cited the popular SIR model, which assumes that the entire population of those affected by the infection is divided into three exclusive categories: the healthy but Susceptible to being infected, the Infected and the Removed, or those who are recovered, vaccinated or died (Figure 1). Since the pandemic is a dynamic system, the number that characterizes these three categories can vary over time and the laws that make it vary are expressed through differential equations, slightly more difficult than those that an average high school student studies when dealing with the topic of derivatives, at the senior year of high school.

It is interesting to remember that *dynamic systems*, are made by all phenomena that change with the passage of time. These are characterized by what we call "time scales" or "characteristic times" t_c. To understand what a characteristic time t_c is, let's take an example.

Let us consider the motion of a projectile launched from a cannon, a very famous dynamic system since the time of Galileo. If we fire with a certain angle and with a certain speed at an instant $t = 0$, we will see the effects of my action after a time t_c which represents the flight time. If I prepare an espresso and put it in a cup, its temperature will tend to change over time. If at instant t = 0 it is $95\,^\circ$C, after a time t_c it will have dropped to about $35\,^\circ$C.

Even for an epidemic there are characteristic times t_c.

For example, if I come into contact with the virus today, how long will it take for me to experience symptoms of the disease? This is generically called the incubation time.

If today there is a mass gathering event, in how many days will the effects on the number of ICU patients be seen? Or, again, suppose that an infected passenger of the omicron variant gets on a plane departing from an African country at time t = 0, with destination Zurich. How long will it take for the first infected person to appear in Switzerland? As it is easy to understand, the study of these characteristic times t_c is very important for the understanding of the epidemic phenomenon and therefore the possibility of finding some effective remedy.

Most importantly, we would like to be able to answer this simple question: when will the pandemic die out? Or what is the characteristic time of the global pandemic phenomenon? Answering is not easy because you must first understand what these t_c times depend on.

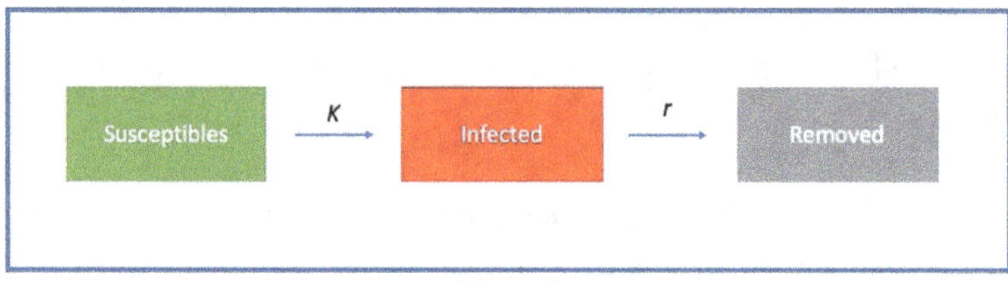

Figure 1

We can say that, once the equations are known, the characteristic times basically depend on the parameters that are part of these equations. In the case of the SIR model there are only two parameters: one regulates the probability of getting infected (K) and one the probability of getting out of the contagion (r) or because of death or cured/vaccinated. There are epidemic models, much more sophisticated, which also have up to 10 different parameters. The determination of the value of these parameters is usually complicated by a hardly negligible reality: the presence of *random fluctuations*.

Let's look a little deeper into the meaning of these *random fluctuations*. In the very simplified epidemic models studied at school, the parameters are usually constant quantities. In reality, however, they vary over time.

Think for example of the parameter K which defines the probability of contracting the disease. This parameter depends first of all on the characteristics of the virus (intrinsic contagiousness), then it depends on the social habits of the subjects. Both of these factors can change over time. The virus, as we have seen, can mutate and thus its contagiousness can change. Subsequent mutations of the virus are generally always characterized by greater contagiousness, because in fact, it is the versions of the virus that take over and end up becoming numerically dominant over a population. Thus, it is the version of the virus that is more effective in reproducing itself.

For this reason, the fearsome Delta variant (the so-called *Indian variant*) which has supplanted the Beta variant (so-called *Brazilian variant*) which in turn has supplanted the Beta variant (*South African variant*) is certainly the most contagious one currently. Now a new variant has arrived, *Omicron*: we will see if it can supplant the Delta and thus show its greater degree of contagiousness.

Social habits can also change. There may be a lockdown, or there may be a community event such as a holiday or wedding. Then there are the climatic conditions which, in the presence of cold climates, lead people to socialize more "closely" and in closed rooms. In short, as you can see, the reasons for the variability of these parameters can be many and when you are faced with a situation with many reasons that you cannot describe all with precision, you rely on a mathematical technique called "adiabatic separation". In other words, we try to highlight the causes of variations that seem to us to be more relevant and that are characterized by a slow speed of variation. For example, the climate and the associated temperature. At this point it is assumed that the parameter K is directly dependent on the local temperature T. But this alone is not enough, because in doing so we would ignore all the other causes. To represent the latter, which are instead sources of fast and independent variations from each other, we can use a random variable, which we will call x. Then we can create an epidemic model where the parameter K is a function both of the slow variable T that we know because it depends on the climate, of the fast (and random) variable x that we do not know, except in its statistical properties.

A dynamic law that has these characteristics was first proposed in the early years of the last century, by the French physicist Paul Langevin to describe Brownian motion (see essay 15 *Much noise about nothing?*) and today we use it to describe the evolution of the pandemic and its characteristic times t_c.

Appeared in Italian in N. 49/2021 *L'Osservatore magazine*.

Brief scientific biography of the author

Luca Gammaitoni, is Professor of Experimental Physics at the University of Perugia, in Italy and the director of the Noise in Physical Systems (NiPS) Laboratory. He is also the founder of Wisepower srl a university spin-off company.

He Graduated at the University of Perugia and obtained the PhD in Physics from the University of Pisa in 1990. Since then he has developed a wide international experience with collaborations both in Europe, Japan and the USA. His scientific interests span from noise phenomena in physical systems to non-equilibrium thermodynamics and energy transformations at micro and nanoscale, including the Physics of computing. In 2016 he has been awarded the Special Breakthrough Prize in Fundamental Physics for the observation of gravitational waves, opening new horizons in astronomy and physics, within the LIGO-Virgo consortium. He authored over 350 papers on top-level scientific journals and few books. He is also the author of 10 patents for industrial applications.

Luca Gammaitoni's scientific interests revolve around the physics of noise. The research topics cover a wide spectrum of topics ranging from the modeling of noise effects on nonlinear dynamic systems (e.g. the case of Stochastic Resonance) to the experimental study of out-of-equilibrium thermal noise, from the study of the behaviour of computing devices at micro and nano-scales (zero-power computing) to the phenomena of transformation of electric energy (energy harvesting).

Among his publications destined for a non-specialist public, the recent "Perchè è difficile predire il futuro" (Why it is difficult to predict the future) written with A. Vulpiani for Dedalo ed. (1919).

From the same author

Sulla inevitabilità del tempo ed altri accidenti
Divagazioni vagamente scientifiche per persone curiose
Luca Gammaitoni
Kindle and in print.

From the same author

Perché è difficile prevedere il futuro
Il sogno più sfuggente dell'uomo sotto la lente della fisica
L. Gammaitoni, A. Vulpiani
2019 Dedalo ed.

The Physics of Computing
L. Gammaitoni,
2022 Springer ed.

www.ingramcontent.com/pod-product-compliance
Lightning Source LLC
Chambersburg PA
CBHW071528150726
48000CB00002B/726